D0118561

Solar Supercapacitor Applications

Ten Hands-On Photovoltaic Projects

by Phillip Hurley

ISBN-13: 978-0-9984727-1-3

Wheelock Mountain Publications
is an imprint of
Good Idea Creative Services
Wheelock VT
USA

Wheelock Mountain Publications is an imprint of:

Good Idea Creative Services
324 Minister Hill Road
Wheelock VT 05851 USA

ISBN-13: 978-0-9984727-1-3

Library of Congress Control Number: 2017931842

Library of Congress subject headings:

Photovoltaic power systems--Handbooks, manuals, etc.
Photovoltaic power systems--Amateurs' manuals.

Disclaimer and Warning

The reader of this book assumes complete personal responsibility for the use or misuse of the information contained in this book. The information in this book may not conform to the reader's local safety standards. It is the reader's responsibility to adjust this material to conform to all applicable safety standards after conferring with knowledgeable experts in regard to the application of any of the material given in this book. The publisher and author assume no liability for the use of the material in this book as it is for informational purposes only.

Contents

Solar Supercapacitor Power Supplies

Supercapacitor basics

Supercapacitors, also known as ultracapacitors, are similar to conventional capacitors in that they are energy storage devices. They store energy in an electric field during charging from a direct current (DC) power source such as a PV panel. However, they differ from conventional capacitors in that they have much higher capacitance capabilities.

Capacitance is defined in terms of charge storage and is designated in units called farads. Charge storage is affected by three basic factors in a capacitor. These are plate/electrode area, plate/electrode spacing, and the dielectric material used for separators. "Super" or "ultra" capacitance is attained, for the most part, by using electrode materials with much larger surface area per volume than conventional capacitors, although electrode spacing and separator material also play a role.

Technically, supercapacitors are known as electrochemical double layer capacitors, or EDLCs. The electric double layer differs from a conventional capacitor in that the dielectric is nanometers thin, which also contributes to creating high capacitance. The nature of the structure of the double layer, however, limits the voltage to 2.5 or 2.7 volts. At higher voltages the electric double layer breaks down, causing capacitor failure. At the present time researchers are working to resolve the voltage limitation barrier.

In summary, electrochemical double layer supercapacitors consist of porous, high surface area positive and negative electrodes, a nanometers thin separator with all three being immersed in a liquid electrolyte which is usually composed of acetonitrile and ionic salts. The materials which give supercapacitor electrodes their unique capabilities are currently either carbon aerogels, activated carbon, carbon nanotubes, or conductive polymers with extremely high surface area characteristics.

Supercapacitors vs batteries

In comparison with batteries, supercapacitors have lower energy densities but their power density is greater. Power density is a combination of the energy density and the speed that the energy can be drawn out of the battery or supercapacitor.

Batteries have much slower charge and discharge times. Supercapacitors have a time constant of between one and two seconds. This means that you can charge a supercapacitor to 63.5% of its capacity in 1-2 seconds. A capacitor is considered fully charged after five time constants. Thus, you can fully charge a supercapacitor within five to ten seconds, and fully discharge a capacitor within the same amount of time if that is what you want to do.

Supercapacitor charge and discharge times are only limited by the heating of the electrodes, whereas batteries depend on the slower movement of charge carriers in the electrolyte. That being said, supercapacitors lose voltage quickly while in use, whereas batteries will maintain voltage for a longer period of time. Unlike batteries, supercapacitors can be totally discharged to 0 volts with no harm.

Supercapacitors also have much better temperature tolerance than batteries and will operate well from -40°C to +65°C.

Supercapacitors have a much longer life cycle than batteries. Life cycles vary by brand from 100,000 to 1,000,000 cycles of charge and discharge.

Suitable applications

Basically supercapacitors are best suited for short duration, pulse power, or longer duration, low current draw situations.

Devices that can operate within a wide voltage range are best suited to being powered by supercapacitors. For instance, a 2.5 volt supercapacitor will be able to provide about 75% of its stored energy if the load can operate in a voltage range of 2.5 volts to 1.5 volts.

Design options are available to suit most power needs. In most instances, the photovoltaic panels are always connected to the supercapacitors and constantly charging during daylight conditions. In this manner a more or less steady application of voltage and current can be maintained. For nocturnal applications, the supercapacitor bank has to be designed with enough capacitance to meet power requirements until daylight. Beyond this, a system could be designed to integrate rechargeable batteries so that the system can ride through a number of limiting conditions.

System components

Solar supercapacitor systems can either be dedicated and device specific, or multiple task universal power supply systems.

Simple solar supercapacitor power supplies consist of:

◆ Photovoltaic panels

◆ Peripheral components such as voltage and current meters and a variety of outlets, connectors and indicator lights

◆ A single supercapacitor or a supercapacitor bank

◆ A supercapacitor bank may or may not include balancing resistors, diodes, voltage balancing circuitry

◆ Optional components include:

    ◆ voltage regulator and/or blocking diode

    ◆ rechargeable battery or battery bank

    ◆ output voltage and/or current regulation

    ◆ micro controller, wireless circuitry and/or other types of sensor activated controls

Design considerations

Key design considerations for a solar supercapacitor system are:

◆ Current and voltage needed by the load

◆ Frequency and duration of duty cycle

◆ Ratio of night/daylight use

◆ Amount of sun available (solar insolation) on a daily basis

◆ Basic environmental factors such as whether the system will be portable or stationary

Whatever the specific parameters, it is important to design for simplicity, economy, longevity and ruggedness.

Hybrid battery supercapacitor systems

Coupled in parallel with batteries, supercapacitors provide an excellent buffer for the battery or battery bank when current surges are needed to start electromagnetic devices such as motors. In this combination, supercapacitors can significantly add to the lifetime of a battery or battery bank. This makes them extremely useful additions to solar or other types of renewable energy power supplies. Although not a proven fact and something that has to be tested, the author is of the opinion that the average battery or battery bank life can be extended by up to at least 30% with the addition of supercapacitors to the system. This would mean that for a deep cycle battery with a life cycle of 5 years in a photovoltaic system, supercapacitors added to the batteries would give a life cycle of 6$\frac{1}{2}$ years.

A hybrid system can use fewer batteries or a smaller battery than an all battery system for the same application, when the application requires surges of power. The supercapacitors can provide for the needed surges, so fewer amp hours are needed from the battery or battery bank.

Range for available supercapacitors

At present EDLCs are available as 2.5 volt or 2.7 volt cells. They are not current sensitive, so you can pump as many amps into them as you want. They are, however, voltage sensitive so it is generally advised not to charge 2.5 volt cells beyond 2.6 volts per cell and not to charge 2.7 volt cells beyond 2.8 volts per cell. Always follow the charging parameters listed on the technical specifications sheet for your particular brand and model supercapacitor.

Supercapacitors are available in a wide range of capacities. Maxwell, for instance, provides the HC power series in 2.7 volt with a range of 25 to 150 farads. Their PC 2.5 volt series have a 10 farad capacitance, their BC 2.5 volt series (as of this writing) has a range of 140 to 350 farads; and their MC 2.7 volt series has a range of 650 to 3000 farads. The BC, HC and PC series have a charge and discharge duty cycle rating of 500,000 cycles. The MC series has a duty cycle rating of 1,000,000.

Terminal configurations vary with model and manufacturer and range from threaded terminals, quick connect, and button, to two pin radial lead, or solid weldable terminals.

Supercapacitor arrangement

A supercapacitor system can consist of one supercapacitor, or two or more supercapacitors connected in series, parallel, or combinations of parallel and series.

Single supercapacitor

Many systems only require one supercapacitor, and the rating of the supercapacitor is what will be available to the load.

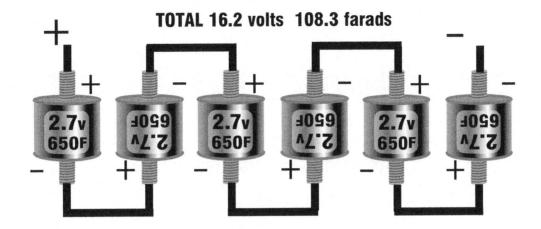

Supercapacitors connected in series

Series connected banks

If you wish to augment the voltage delivered to the load you can arrange the supercapacitors in series by connecting the negative and positive terminals, which will add the voltage of each supercapacitor for final output.

Presently, EDLCs are rated at 2.5 volts or 2.7 volts per cell. Adding cells in series will result in set voltage configuration options. For instance, two cells will give you 5/5.4 volts; 3 cells - 7.5/8.1 volts; 4 cells - 10/10.8 volts; 5 cells - 12.5/13.5 volts; 6 cells - 15/16.2 volts, and so on. There is no limit to the voltage addition.

However, you pay for the voltage addition, because connecting supercapacitors in series will diminish the capacity of the total configuration to a value that is less than the capacity of one of the supercapacitors in the bank.

The degree of reduction of capacity depends upon how many supercapacitors are connected together in series. If you connect two 650 Farad supercapacitors in series, you will reduce the charge capacity to one half the value of the capacitance rating, or 325 farads. Each additional series connected supercapacitor will reduce the charge capacity of the bank further. If you connect six 650 farad supercapacitors in series you would only get $\frac{1}{6}$ of the charge capacitance of one cell, or around 108.3 farads.

For some applications this loss of capacitance is not of importance. However, for other applications, more capacitance may be required at these higher voltages. In that case you can use a parallel series arrangement which connects two or more cells in parallel and then connects these parallel connected capacitors to a similar parallel connected string in series. For example, if you connected twelve 650 farad, 2.7 volt

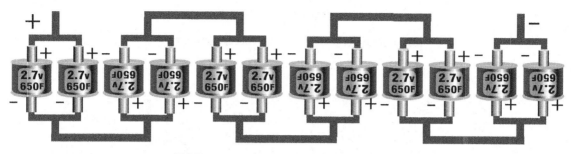

TOTAL 16.2 volts 216.6 farads

Parallel connected pairs of supercapacitors connected in series

supercapacitors in parallel-series as in the illustration above, you would get 16.2 volts, at around 216.6 farad capacity.

You can, of course add more capacitors in your parallel string to augment capacity but it begins to get expensive. There are other less expensive ways to increase voltage using DC to DC converters with a parallel string.

Parallel configurations will add the capacitance of each cell but the voltage remains the same as one supercapacitor. For example, if you connect six 2.7 volt, 650 farad supercapacitors in parallel you would get 2.7 volts at 3900 farads. This is the most efficient configuration. The capacity of each cell is used fully, however the limitation of 2.7 volts can be a problem for powering equipment that requires higher voltage.

TOTAL 2.7 volts 3900 farads

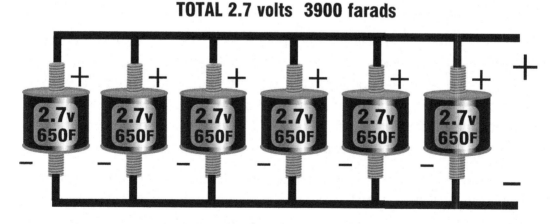

Supercapacitors connected in parallel

DC to DC converters

To overcome this low voltage limitation, DC to DC converters are used with parallel strings to boost voltage to needed levels. DC converters, however, do have a conversion loss factor that must be considered in application.

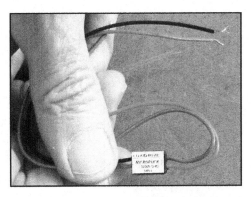

Micro Puck boost converter

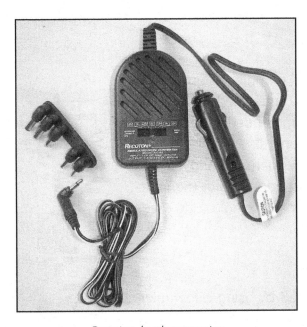

Recoton buck converter

There is a wide variety of DC to DC converters available, but not all will be suitable for your particular supercapacitor applications. They can be used either to raise voltage (boost) or lower voltage (buck), or a combination of both to reach your desired ends. The primary consideration in choosing an inverter is that it is rated to deliver the amount of current and particular voltage needed for your load. Several projects in this book were designed to give an example of converter use with supercapacitors.

Voltage balancing in series connected capacitors

When supercapacitors are connected in series, any given cell in the string will not always retain the same voltage as other cells in the string. This is due to a minor difference in leakage current. When recharging a group of series connected cells, some may be at different voltage than others. This can create a situation where some cells become overcharged during the charging process. Overcharging can reduce supercapacitor life expectancy by damaging the cell. To avoid this situation either an active or passive balancing system is used to address the varying voltage in the string cells.

Active Balancing

Active balancing employs balancing circuitry to maintain an equal voltage on all capacitors connected in series. This is the most efficient way to maintain an equal voltage on cells.

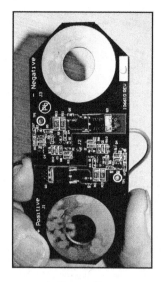

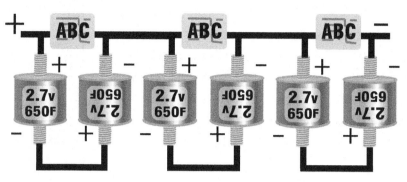

Series connected supercaps with active balancing circuit

Active voltage balancing circuit board

Maxwell offers an integration/voltage balancing kit that can be used for their MC series and is recommended for very high duty cycle operations. It is relatively inexpensive but more expensive than passive balancing with inexpensive resistors. Most supercapacitor manufacturers will supply an application sheet with an active balancing circuit schematic if you wish to build your own.

Passive balancing

Passive balancing is usually accomplished by using a low tolerance resistor connected to the positive and negative terminal of each capacitor. A resistor used in this manner is termed a bypass resistor. The purpose of the bypass resistor is to offer a controlled path of less resistance that will dominate the leakage process in each cell making the process more uniform by equalizing all cells similarly with the same value resistive path. The value of the bypass resistor you select will depend on the duty cycle that you intend to put your capacitors through.

If the capacitors must recoup quickly you will have to put a higher-wattage lower-value resistor between the terminals so that they can balance quickly. If intended use is more or less occasional you can use a higher-value lower-wattage resistor.

Basically a lower value resistance will reduce the time to balance but increase losses. A higher value

Series connected supercaps with resistors

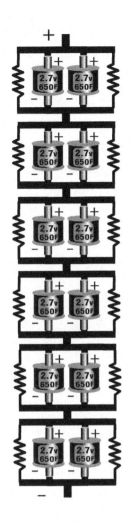

Parallel connected pairs, pairs connected in series, bypass resistors included .

Total 16.2 volts, 216.6 farads

resistance will increase the time to balance but decrease losses. So, there is a trade off to be reckoned with and the final decision can only be made with reference to the duty cycle needed.

In regard to duty cycle you will have to either experiment to determine the most advantageous resistance for your application, or ask the technical staff of the manufacturer or supplier.

To determine the value of the balancing resistor needed for your application, you need to know the leakage current for that particular supercapacitor model. The leakage current for your supercapacitor is listed in the technical specification sheets provided by the manufacturer for your particular brand and model.

For example, Maxwell 140 Farad 2.5 volt supercapacitors have a leakage current of 0.1 milliamps, 150 Farad 2.7 volt supercapacitors have a leakage current of 0.5 milliamps, and 650 Farad 2.7 volt supercapacitors have a leakage current of 1.5 milliamps.

To demonstrate how to determine the appropriate ratio of bypass leakage current to cell leakage current for your applications we will use a 20:1 ratio. With a 20:1 ratio you would multiply the cell leakage current by 20 and divide the rated voltage of the capacitor by this figure. This will give you the resistance needed for a ballpark start. For many applications a 10:1, or 20:1 ratio is used, however you can have a 5:1 or a 100:1 ratio, and so on.

The idea is to apply a bypass resistor that will have less resistance then the cell resistance so that the resistor will dominate the leakage process. Each capacitor, of course, would have the same value of resistor added to it so that the leakage will be uniform with all capacitors, rather than having the random variation cell to cell that would occur without the resistors. You should use low tolerance resistors rated at 1%. What are termed "precision resistors" can be used for tighter tolerances below 1%.

One of the best ways to get to know what a resistor supercapacitor combination will do for your particular application is to charge two or more supercapacitors. Leave one supercapacitor to self discharge with no resistor attached across its terminals. With another supercap, attach a 275 ohm 1%, 0.25 or 0.5 watt resistor. Compare the discharge times.

To calculate the power rating of the resistor needed for your application, divide the voltage by the resistance value. This will give you the current flow through the resistor. Multiply the current flow by the voltage to get the minimum watt rating needed.

Photovoltaic panels and cells

There are many types of PV cells available for those who want to build a panel with specific outputs to match a supercapacitor application. The most commonly used cells at the moment are silicon polycrystalline, monocrystalline and amorphous cells.

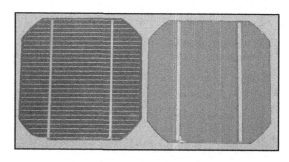

Monocrystalline photovoltaic cell, above.

Below left, flexible amorphous cell; right, polycrystalline photovoltaic cell

For most solar supercapacitor applications, high current output PV cells are important as they reduce charging time. Per area of PV cell, monocrystalline has the highest current output, polycrystalline comes in second and amorphous silicon is third.

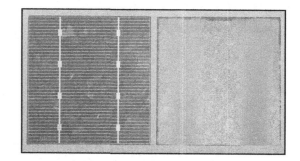

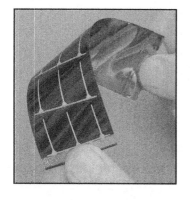

When to use amorphous PV cells

Generally, the best choice for quick charging would be mono- or poly- crystalline cells. However, there are factors that could make silicon amorphous more desirable. In some locations cloud cover is more prevalent than not. Amorphous cells react more favorably to diffuse light conditions and can produce current in conditions that poly- or mono- crystal cells would under perform.

Amorphous PV cells, while commonly available with a rigid substrate, can also be put onto a flexible polymer substrate. These flexible PV cells are great for a variety of applications where wafer type cells would be stressed because of their rigidity. For instance, flexible polymer amorphous cells can be mounted on clothing and fabric. They are favored for model planes, rockets and other devices that do not need much current, but where flexibility is important.

Flexible amorphous silicon cells are commonly available in 3, 3.6, 4.2, 4.8, 6, 7.2 and 15.4 volt cell strip configurations. You can, of course, connect these in series, parallel, or series parallel to create the custom voltage and current desired.

There are amorphous cells that are not based on silicon technology such as copper indium diselenide. These produce more current per area than amorphous silicon but they are at present considerably more expensive.

Silicon wafer cells

Essentially monocrystalline cells are the best choice for charging supercapacitors because of their higher current output per area of cell, compared to other types of cells.

Mono- and poly- crystalline cells are rated at around .5 volt per cell. The exact voltage of a cell will vary somewhat above and below .5 volts depending on the type of cell. Most cells will run a little over .5 volts. Generally, the larger the surface area of the cell, the greater the current will be, compared to smaller cells of the same type from the same manufacturer, and made by the same process. If you compare cells from different manufacturers, an area comparison might not be accurate for current output.

PV panel voltage needs

Please note that what are termed 12 volt system panels in fact deliver more than 12 volts. The reason for this is that panels were originally built to charge 12 volt batteries, and multiple series connected batteries that could provide 24 or 48 volts and so on. 12 volt batteries, in fact, are fully charged at about 12.6 to 13.2 volts per 6 cell battery. It is necessary to have a higher charging voltage than the actual battery voltage to charge the battery, so the power supply (in this case the solar panel or array) must provide a few volts above the battery's actual voltage, usually 13.8 to 14.7 volts for a 12 volt battery.

A blocking diode is needed to prevent reverse current loss at night as the batteries tend to discharge back into the panels. A blocking diode has a voltage loss of .3 to .7 volts. Thus, if your system needs a blocking diode, the panels should have enough voltage to make up for this voltage loss.

Other factors that create voltage loss that should be considered are days without much sun, which will cause the output voltage of the panel to drop a bit; and voltage drop from wire runs from the panels to the battery bank.

Basically by the time the electrons make it to the battery there must be enough voltage to charge the battery adequately. For a hybrid supercapacitor and battery system all these considerations are the same as for batteries alone. For a stand alone supercapacitor system the voltage drop considerations are the same, but you do not need a higher charging voltage to charge the supercapacitors, so less voltage is necessary from the panels.

As an example, a 15 volt supercapacitor bank consisting of six 2.5 volt supercaps in series needs 15 volts to charge it fully. The solar panel can be comprised of from 31 to 32 cells, 4 to 5 fewer cells than if a battery were involved in the system.

If you purchase your panels for a supercapacitor power supply system, there are many kinds of panels available for 12, 24, and 48 volt configurations. Finding high current outputs for other voltages is a bit harder and you will probably have to construct your own panel for 2.5 and 2.7 volt output systems.

Photovoltaic connections

◆ Connecting PV cells, strings, and panels in series adds voltage but not current.

◆ Connecting PV cells, strings, and panels in parallel adds current but not voltage.

In most commercial PV panels, the individual cells are connected to each other in series, meaning that the negative top (face side) of each cell is connected to the positive back of the next cell to form one string of cells, so the

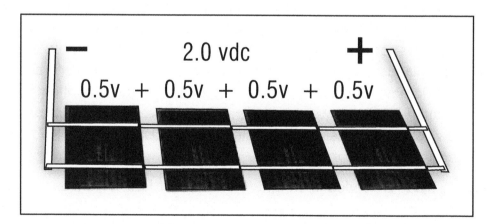

A string of PV cells connected in series. The back side (positive) of each cell is connected to the face side (negative) of the next cell in the string

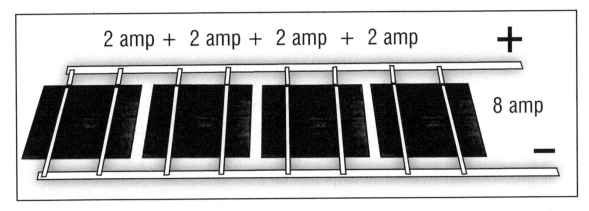

A string of PV cells connected in parallel. The back side (positive) of each cell is connected to a positive bus, and the face side (negative) is connected to a negative bus.

panel voltage is the sum of the voltage of each cell in the panel, and the panel amperage is the same as the current output of one of the cells.

If the PV cells were parallel connected in a string, the back side of each cell would be connected to a positive bus and the face side of each cell would be connected to a negative bus. The panel current output would be the sum of each cell's current output, and the panel voltage would be the same as the voltage of one cell.

As an example, with thirty-six 4 amp cells, you could series connect all the cells to produce an 18 volt panel that will deliver around 4 amps of current. If these same cells were connected to each other in parallel, the panel would be .5 volts at 144 amps.

A panel layout in which the cells within the strings are connected to each other in parallel, and the four strings are connected to each other in series.

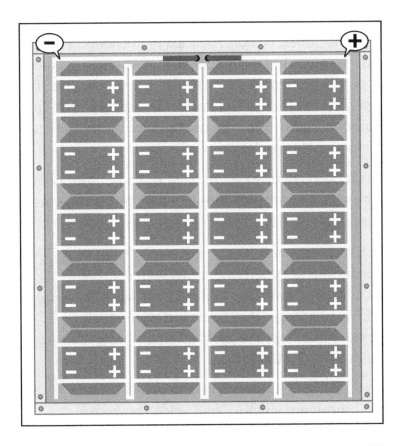

You can also connect cells, strings or panels in series-parallel combinations to boost the current output. For instance, if you have two 36 cell strings of series connected cells; and connect the two strings to each other in parallel, the resulting panel of 72 cells will have an output of 18 volts at 8 amps (assuming 4 amp cells). Another way to get the same results would be to connect two 12 volt system panels in parallel. The same two 12 volt system panels could be connected in series for an array that will deliver 24 volts at 4 amps.

So, there are several different ways that PV cells, strings and panels can be configured to give the specific voltage and amperage needed for any particular system. Obviously, if you build your own PV panels, you can tailor the power supply very specifically for your application.

For more information about constructing solar panels, see *Build Your Own Solar Panels* and *Build a Solar Hydrogen Fuel Cell System*.

Charge controllers and voltage regulators

Most photovoltaic battery systems have a charge controller that regulates the voltage from the panel to the battery. This insures that the batteries do not overcharge and gas excessively.

Some panels do not produce a high enough voltage to create problems with excessive gassing. They deliver between 13-15 volts and are often called battery maintainers or self regulating panels. With this type of panel a charge controller is not necessary, just a diode. This type of lower voltage panel can be used for stand alone supercapacitor systems. They are not frequently used for supercapacitor and battery hybrid systems, although they can be.

Charge controllers (or at least a diode used for voltage regulation) must be used with solar supercapacitor and battery hybrid systems unless the voltage coming from the panels is so low that it can not overcharge the battery.

Charge controllers can be constructed or purchased. Commercial charge controllers are most readily available in the 12 volt, 24 volt, and 48 volt range. They come in a wide variety of current ratings, the most common being a range of 10 to 60 amps.

Most charge controllers allow you to set the charging voltage and also have an equalization charge switch. Generally, charge controllers are set to charge flooded lead acid batteries at around 14.5 volts. Sealed lead acid batteries are charged at about 13.8 to 14.1 volts.

The equalization charge switch allows about 15.5 volts to the battery. An equalization charge is only performed on flooded lead acid batteries if the cells get out of voltage balance with each other. Keeping the voltage in balance greatly lengthens the lifetime

Charge controller constructed from a kit

of the battery. However, you need not and should never equalize the cells of a sealed lead acid battery as it will destroy the battery.

A system powered with a 12 volt panel should have a 12 volt regulator. The amp rating for the controller will depend on the current available from the PV panel or array. For instance, if you have one 12 volt 4 amp panel then you will only need a 10 amp controller. If you have three of these same panels connected in parallel you will need a 20 amp controller. Most charge controllers have a blocking diode that makes it unnecessary to add a diode to the circuit.

Either a charge controller, Schottky diodes or general purpose rectifier diodes can be used for voltage regulation for solar supercapacitor stand alone systems.

Batteries for hybrid systems

Most supercap and battery hybrid systems use flooded lead acid batteries as they are more forgiving and last longer than SLA (sealed lead acid batteries). The downside is that flooded lead acid batteries require maintenance to check water levels occasionally whereas sealed batteries do not require such maintenance. If you use sealed lead acid batteries in a system, be sure your charging voltages do not exceed limits stated by the manufacturer. Sealed lead acid batteries deteriorate rapidly when overcharged.

Any type of rechargeable battery can be used with a photovoltaic supercapacitor system. Deep cycle flooded lead acid are the most common type for home power and industrial systems. Deep cycle sealed lead acid are also used but not as frequently. Lithium, metal hydride and nickel cadmium are used for smaller applications. Edison nickel iron alkaline cells are another choice, but they are not frequently used.

Lead acid cells and batteries come in 2 volt, 4 volt, 6 volt, 8 volt, and 12 volt packages. Other voltages are available in single package design, but they are not as common.

2 volt deep cycle lead acid batteries such as the Trojan L16RE-2V, Rolls Surrette 2 volt cells, or similar are the best choice as you can fix a supercapacitor across the terminals of each 2 volt battery or cell. Six of these batteries or cells connected in series will make a 12 volt cell battery supercapacitor combination that does not require balancing resistors or an active balancing

2 volt battery cell connected in parallel with a supercapacitor

circuit. Most 2 volt cells for home or industrial use are quite large and expensive, though there are also smaller 2 volt cells available, such as the Cyclon.

12 volt battery bank

By far the most popular lead acid battery for home power systems is the Trojan T-105-RE, a 6 volt deep cycle flooded lead acid battery. These batteries are connected in series to create 12, 24, or 48 volt systems. They are popular because of their relative affordability, availability and quality.

Each battery or individual cell has its own specific charging parameters.

For instance, Hawker cyclon sealed lead acid 2 volt cells are fully charged at 2.14 volts. The manufacturer recommends charging at about 2.45 to 2.50 volts per cell for cyclic applications and 2.25 to 2.30 volts per cell for float charging. This means that a six cell, series connected 12 volt bank of these particular cells are fully charged at around 12.84 volts. If the cells are used on a daily basis, a 12 volt series bank should be charged at 14.7 to 15 volts. If the cells are infrequently used, such as for backup power, then they should be float charged at around 13.5 to 13.8 volts. These particular cells should not be discharged beyond 1.93 volts for each cell.

As a general rule for 12 volt flooded lead acid battery systems, I charge at around 14.5 volts (2.4 volts per cell); and for sealed lead acid around 13.8 to 14.1 volts per 12 volt system (2.3 to 2.35 volts per cell). These are the parameters I use if I lack specific charging information for a particular battery.

Lead acid battery systems will degrade if undercharged and/or overcharged consistently. The life cycle of the battery is also affected by the depth of discharge on a consistent basis. Deep discharges on a daily basis will tend to shorten the lifetime of the battery or cell.

Design considerations for a hybrid system

Adding supercapacitors to a photovoltaic battery system is pretty straightforward, and there are only a few considerations to note.

The voltage to the supercapacitor and battery bank cannot exceed the rated voltage of the supercapacitors. For instance, for a 12 volt system that is charging flooded lead acid batteries, you would need a bank of series connected supercapacitors rated at 15.5 volts per bank. This is because general charging in a flooded lead acid battery system will be at around 14.5 volts with an occasional equalizing charge voltage of 15.5.

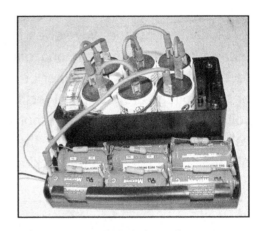

*Series connected
12 volt hybrid system*

*Parallel connected
2.0 volt hybrid system*

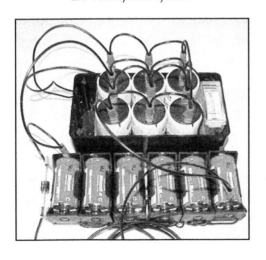

Supercapacitors are presently available in two voltage ratings, 2.5, and 2.7 volts per cell. A bank of six series connected 2.5 volt supercapacitors will give you a total voltage of 15 volts. A bank of six series connected 2.7 volt supercapacitors will give you a total voltage of 16.2 volts. At first glance it would seem that you need to use the 2.7 volt bank at 16.2 volts to cover the 15.5 volts needed for the system. This would be a good choice; however, you can also use the 2.5 volt supercapacitors as they have a overcharge margin of about .1 per cell. This would allow you about 15.6 volts for charging. This is just within the range of general charging and equalization voltage needed.

The 2.5 volt bank will only receive a charging voltage of 14.5 most of the time, with an occasional charge during equalization for a few hours at 15.5 volts. The equalization voltage on charge controllers can vary as to their set voltage and they may be subject to voltage drift. Be sure not to surpass the limits for the supercapacitor. Test the controller with the batteries first, then connect the supercapacitors when you are sure there is a good match. It is up to you whether to apply a more liberal interpretation to charge voltage, or remain conservative in your approach.

Batteries must be fully charged. They are degraded by charging to a lower voltage. Supercapacitors, however, are not affected by undercharging.

Diodes

Diodes act as one way valves for electrons. They allow electrons to flow in one direction and block reverse flow. Diodes can either be low voltage drop or high voltage drop. General rectifier diodes have a voltage drop of about .7 volt, Schottky diodes have a voltage drop of about .3 volts. These characteristics allow diodes to be used as blocking diodes from panels to supercapacitors or hybrid

General rectifier diodes

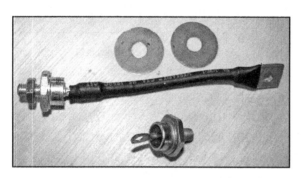

*Above, high current diode;
below, diode connected to heat sink*

supercapacitor and battery systems to prevent losses from these storage devices back into the panel when current is not flowing from the panel to the system. Diodes are also used as simple voltage regulators singly or connected in series to reduce voltage from panels to more exactly match the solar panel output to the supercapacitor's rated voltage.

Diodes are rated for their voltage and current carrying capabilities. It is common practice to use diodes rated for higher voltage and current than necessary. For instance, if your panels output 18 volts at 4 amps, any diode rated above those designations would be sufficient. For this example, a 40 to 50 volt, 10 amp diode would commonly be used, although lower values can be used. Certain values are more readily available and this will affect your choice. The main concern is not to use a value that is lower or right at the edge.

Some diodes, when they reach their breakdown voltage rating, allow the current to flow through them. These are called Zener diodes and can be used instead of balancing resistors. When a certain voltage is reached, the current flows through the diode and thus maintains the voltage of the capacitors at the rated voltage of the Zener. If used for balancing you will need to use 2.5 or 2.7 volt low tolerance Zeners. Zeners can also be used for voltage regulation.

Diodes come in a variety of shapes and sizes. Higher current diodes are generally heat-sinked to dissipate heat.

Resistors

Resistors are used to limit current. They are very useful when connected in parallel to each capacitor cell to dominate leakage current and thus balance the voltage of one cell with another in a series string connection. They can also be configured for use as voltage dividers to limit voltage to a supercapacitor supply.

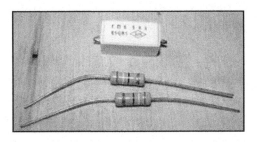

Resistors

Resistors come in a variety of tolerances. The best option for accuracy is to use 1% tolerance resistors or precision resistors for tighter tolerances, if so desired. Resistors also have a variety of watt ratings. It is important that any resistor used in a circuit can dissipate the heat produced by the resistance to the current flow produced in the resistor, thus the proper watt rating is needed.

Cable, wire, connectors, and fuses

As with any electronic or electrical circuit, you need to understand some basic electrical formulas, voltage and current ratings of cable, wire, connectors, and fuses so you can safely and effectively construct and connect various components of a solar supercapacitor system. If you are not acquainted with the basic construction of photovoltaic systems *Solar II* has this information with pertinent formulas and tables.

Generally, 12 gauge wire with a current carrying capacity of 23 amps, to 10 gauge wire with a current carrying capacity of 32.5 amps is used from the solar panels to

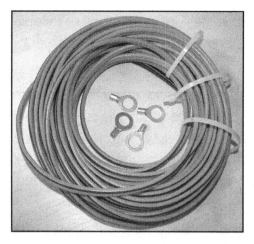

Zip cord and terminal rings

the charge controller and/or supercapacitor or combination supercapacitor and battery supply. This is usually UFB, zip cord or tray cable. Of course you can use smaller diameter wire if the current delivery is below these figures.

Supercapacitor to battery connections can be anywhere from 6 gauge to 4/0 depending on your current draw needs. The same rule applies for connecting stand alone supercapacitors, and supercapacitor/battery hybrid banks, to the load or inverter. Wire size is dependent on current draw needs.

In general, battery and inverter cable is used for DC connections and NM or NMC is used to carry the 120 volt from the inverter to outlet in most stationary home power systems.

There are many types of connectors that can be used, such as lug, spade and ring terminal, quick connects and so forth. When attaching terminal connectors to wire it is important to crimp firmly so that the connection does not

Battery cable

move and is tight and secure. Crimped sections should be covered with shrink tube, electrical tape, etc., for good electrical insulation. Terminals and connectors must be rated for current and voltage.

Fuses provide circuit protection from overload and short circuit current. All positive wires from a supercapacitor bank, or a hybrid supercapacitor and battery bank, need to be fused. The voltage rating of the fuse should be equal to or greater than the system voltage, and the ampere rating of the fuse should not exceed the current carrying capacity of the circuit. A good rule of thumb is to use a fuse rated for 125% to 150% over the amp flow expected within the circuit under normal operating conditions, and no higher than the ampacity of the smallest wire in the circuit. For circuits with motors, consider the heavy current surge that a motor creates briefly on startup.

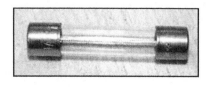

Fuse

A higher current rated fuse or a slow blow or time delay fuse can answer to these issues, depending on your needs. For most circuits a fuse is advisable and necessary, though for some, such as welders and ignition devices, fuses may not be used.

Inverters

For applications that require the use of AC, an inverter is needed in the system. Inverters convert DC (direct current) to AC (alternating current). This makes it possible to use AC appliances with DC supercapacitor, or supercapacitor and battery power supplies. Inverters are available for portable as well as stationary applications.

An inverter's watt rating must match the power requirements of the devices you will be using. Inverters can be used with a stand alone supercapacitor bank; however, many inverters have an over voltage shutdown which is factory set at 15.5 or 15 volts depending on the inverter.

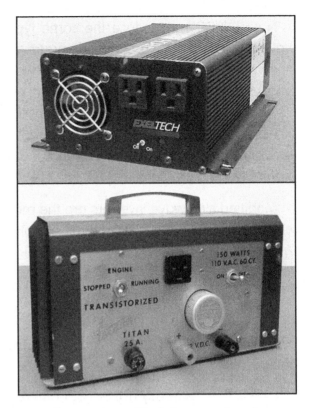

Above top, modified sine wave inverter; above bottom left, square wave inverter; below, portable (very small) car accessory inverter;

If you use a bank of six 2.5 volt series connected supercapacitors you will be fine. If you use six 2.7 volt series connected supercapacitors, you can either just charge the supercaps to 15 volts; or charge to full 16.2 volts if you use an inverter that operates with a higher cutoff voltage. Invertors such as the PowerStar UPG 400, 700, and 1300 have a high voltage cutoff around 16.8 volts which makes them compatible with fully charging a six cell 2.7 volt series connected supercapacitor bank. To make full use of the capacity of your bank you should ideally charge it to its full charge potential when feasable to insure longer runtimes for your equipment. For hybrid

supercapacitor battery systems, an inverter with a shutdown voltage of 15 volts is not a concern as the output voltage to the inverter is regulated by the battery.

Not all inverters produce the same type of wave form. Some produce a square wave, others produce a modified square wave (sometimes referred to as a modified sine wave), and some produce a sine wave. Most inverters on the market produce what is referred to as a modified sine wave. This is sufficient for most purposes. Usually, pure sine wave inverters are used when electricity is fed back into the grid or when used with appliances that do not work well with modified sine wave inverters. Square wave inverters are rarely used although they will power lights and certain motors with no problem.

Modified sine wave inverters are the most used and less expensive than pure sine wave devices. If you are not acquainted with inverters, more information can be found in *Solar II*.

Electrical safety

When working with 120 volt AC circuits or extreme high voltage circuits, you need an understanding of proper grounding practices in order to avoid electrical shock hazard.

NFPA 70 (specifically article 690) otherwise known as the National Electrical Code (NEC) published by the *NFPA* has the basics for general system requirements for photovoltaic systems. For extreme high voltage applications, refer to relevant available literature for safety considerations and application notes.

If you have a good understanding of basic photovoltaic systems, it is not difficult to properly integrate supercapacitors into your system.

For more information about basic electronic and electrical circuits, I suggest Tony Kuphaldt's *Lessons in Electric Circuits* as one of many good references. Other good resources are photovoltaic supplier catalogs such as those supplied by *New England Solar* and *Backwoods Solar*. These catalogs will acquaint you with a variety of devices and contain much information about PV systems in general.

Solar Panel Construction

Solar panels are easy to construct and, for the projects detailed here, consist of:

- PV cells, either polycrystalline or monocrystalline
- Tab and bus ribbon
- Acrylic bar stock spacers and edge supports
- Plexiglas (acrylic) cover and substrate (base)
- Binding post output connectors

First, select the type of cell you wish to use, and decide how the cell connections will be configured, for instance, series or parallel-series, as discussed in the previous chapter.

Once you know the cell dimensions and the connection configuration, plan the panel layout and figure out the dimensions for the substrate, cover and spacers. An easy way to do this is to lay the cells out on a flat surface. I like to space cells about $\frac{1}{8}$" apart, but this can be adjusted to suit your needs. Make allowances for the edge and intercell spacers, binding post connectors, bus ribbon, as well as the space needed between the cells. For more details about solar panel layout, see *Build Your Own Solar Panel*.

Substrate and cover

In this example I used six 5"x 5" cells. According to my layout calculations, I need two $17\frac{1}{2}$"x $10\frac{1}{2}$" pieces of Plexiglas, and five spacers cut from $\frac{1}{8}$" sq. acrylic bar stock, two $17\frac{1}{2}$" long, two $10\frac{1}{4}$" long, and one $15\frac{1}{2}$" long.

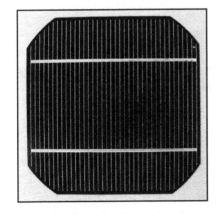 

The 17½" spacers and the 10¼" spacers will be the edges of the panel.

The 15½" spacer will be bonded to the center of the panel. This spacer will support the cover as well as forming a holder for the cell strings. It is shorter than the length of the panel to allow space for the bus ribbon to connect from one string to the next.

Leave the film on the Plexiglas to protect it during cutting.

For this panel I wanted a camera tripod mount attachment, so I also cut a 1½" square from the Plexi sheet for a base to mount the perforated base nut.

Plexiglas sheet comes covered with a thin plastic film. You can leave the film on the piece while cutting to protect the surfaces from abrasion.

The Plexiglas sheet is easily cut by marking the cutting line with a light score, and then deepening the score with a Plexiglas knife (see photos below). Then, align the score with a table edge and apply downward pressure to snap the Plexiglas along the score.

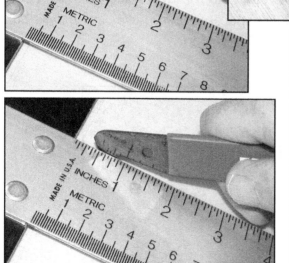

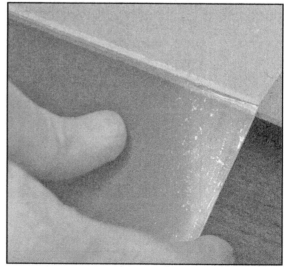

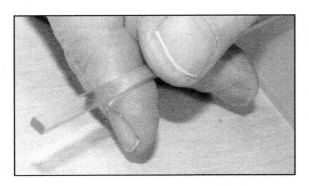
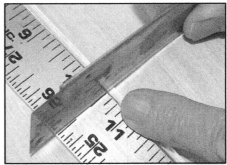

The spacers are cut from $1/8$" square acrylic bar stock with a fine tooth handsaw.

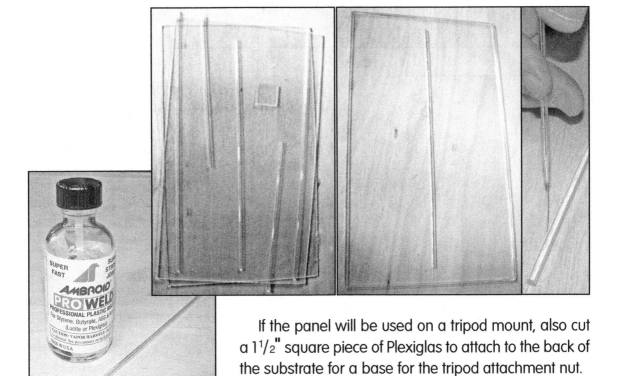

If the panel will be used on a tripod mount, also cut a $1^1/2$" square piece of Plexiglas to attach to the back of the substrate for a base for the tripod attachment nut.

Once you have cut the pieces, line up the spacers on the substrate, secure them and bond them into place with plastic weld.

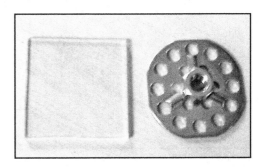

Tripod mount

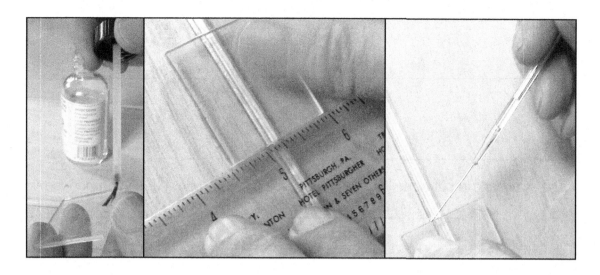

To apply the base for the tripod attachment, coat one side of the square with plastic weld, and press it firmly on the center of the substrate so that it bonds.

You can also apply plastic weld around the edges of the square to ensure a good weld.

Next, coat the $1^1/_2$" square with epoxy and apply the nut to that surface. Let this dry for 24 hours so that it bonds well.

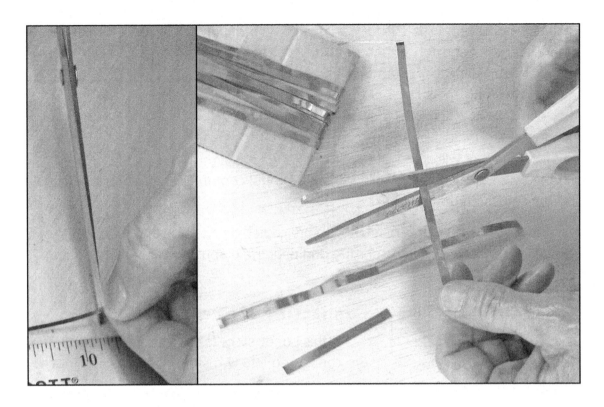

Bus and tab ribbon

Cut the tab and bus ribbon to length. For the example panel we need 16 pieces of tab ribbon cut to 10⅛" long. Each tab ribbon length covers the length of two cells with an added ⅛" between cells.

Two pieces of bus ribbon were cut to 2⅝" and one cut to 7¾". The 7¾" bus ribbon connects the positive output of one string to the negative output of the other string in the panel. The two 2⅝" pieces of bus ribbon are the output leads connected to the binding posts in the panel. Holes should be drilled in the centers of the two 2⅝" bus ribbon pieces to insert the binding post screws through them.

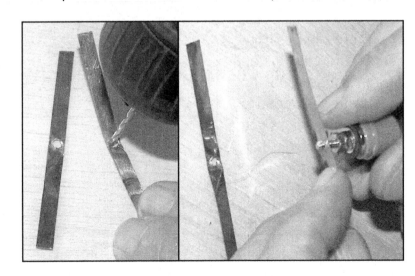

After the tab ribbon is cut to length, coat 12 pieces of the tab ribbon with a thin layer of solder for 5" from one end on one side, and then for 5" from the other end on the other side, leaving ⅛" untinned in the middle of the tab. These are the areas that will come into contact

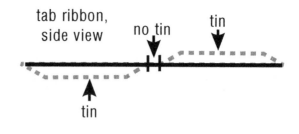

with the cells. This process is called tinning. Do not coat the ⅛" space between cells since this must remain flexible.

The four remaining pieces of tab ribbon are coated on one side only for a length of 5" from one end. These will be the positive and negative output leads for the two strings.

Connecting the cells

For this panel, six cells are soldered in series. They are configured into two strings of three cells each.

For the first soldering operation, lay three cells face side up. (The face side is the negative blue side.) Add flux with a flux pen by running the pen down the fingers of each cell. The layer of flux prepares the surface for soldering and helps ensure a good join.

Next, lay the tab ribbon, tinned surface down, onto the cell fingers and run the soldering iron along the tab. The tinning that you added to the tab will melt as you move the soldering iron along the surface and form a bond with the fingers of the cells.

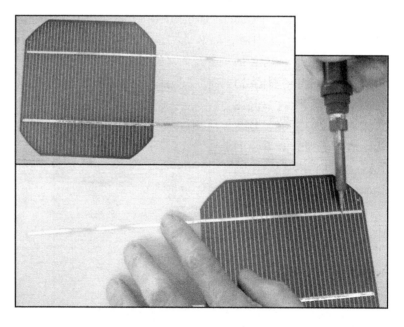

After the tab ribbon is soldered to the fingers on the negative side of the six cells, tab ribbon should be applied to two of the six cells on their back (positive) sides.

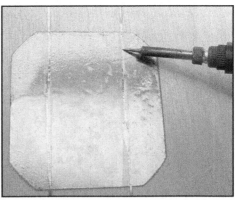

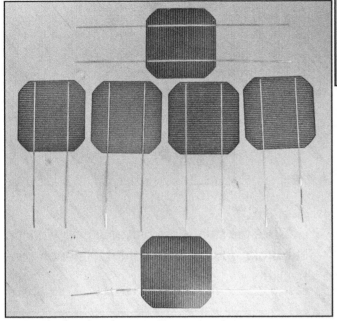

The six tabbed cells are then ready to connect in a series configuration.

The cells will be soldered together into two strings of three cells each, then the two strings will be connected to each other.

It is important to get good alignment between the cells so that the tab ribbon lines up cell to cell and follows the fingers exactly; and so that the cells fit into the panel correctly.

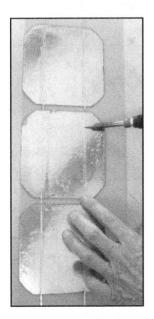

To prepare to connect the cells, align three cells in a straight row with ⅛" space between the cells. You can draw a template to align the cells.

To solder the cells together, lay three of them face (negative) side down, on the template. One of the cells with extra leads should be included in each group of three cells.

When the cells are lined up, roll the soldering iron over the tabs to join the cells. Use a flat square wooden stick to hold the tab ribbon in place while soldering, but do not apply too much pressure because the cells can easily crack.

When you have finished soldering the two strings, place them in position on the substrate.

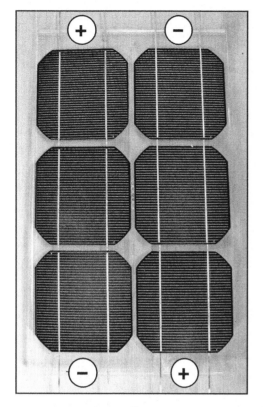

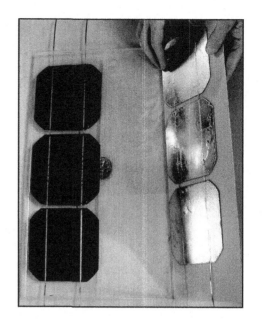

Once the strings are placed properly on the substrate, you can make the bus ribbon connections.

The long 7³/₄" bus ribbon connects the negative output of one string to the positive output of the other string. Lay the bus ribbon over the tab ribbon extensions coming from the cells and note or mark the points where they cross over each other. Tin the bus ribbon at the points where it will connect with the tab ribbon.

One of the 2⁵/₈" bus ribbons is attached to the negative output tabs and the other to the positive outputs tabs coming from the cell string. Lay these two bus ribbons

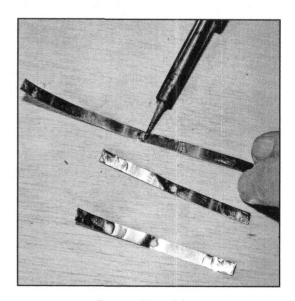

Tinning bus ribbon

across and over the cell tab ribbons and note and mark where they connect. Tin one side of the bus ribbons where they will connect with the tab ribbon and solder the bus to the tab.

When the bus ribbon is tinned, put a piece of thin cardboard under the junctions of the tab and bus ribbons to absorb the heat from the soldering iron. Don't skip the cardboard underlay as the heat from the iron will surely melt the plastic. Lay the bus ribbon tinned side down on the tabs and solder by applying heat to the bus ribbon. After you have done this you can proceed with attaching the two shorter bus ribbons on the other end of the panel.

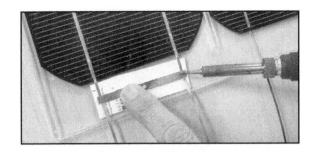

Trim off excess tab ribbon with scissors, as above.

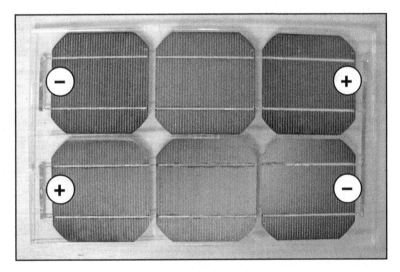

Once the strings are connected to each other, drill a hole in the Plexiglas for the screw terminals from the binding posts. Mark the plastic substrate at the points where the holes in the two $2\frac{5}{8}''$ bus ribbon are (see photo at left). Remove the strings from the substrate.

Drill holes at the two marked points on the substrate. Place the strings back onto the substrate.

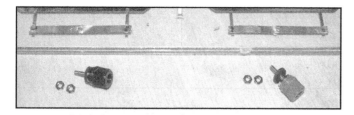

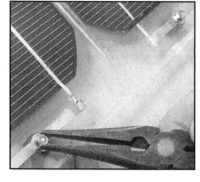

Insert the binding posts through the plastic and bus ribbon and screw on the two nuts that come with each binding post until they are secure.

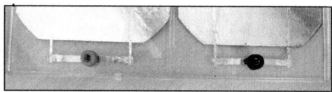

Next, apply some silicone caulk to the top and bottom edges of the cells to hold them in place. You do not need very much silicone, just dab a little along the edges with a toothpick. Smooth out the silicone by running your finger along the edge. Let the silicone dry for 24 hours.

Lay the cover on and apply plastic weld to the edges where the cover meets the edge spacers.

Apply pressure with your fingers as you apply the weld to make sure the pieces are bonding, or use mini clamps to hold the pieces in place while you apply the weld.

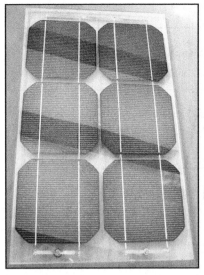
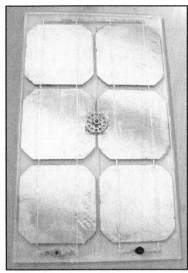

There are many ways to make solar panels. If you are interested in more detailed instructions and variations, see *Build Your Own Solar Panel.*

Two other panel designs:

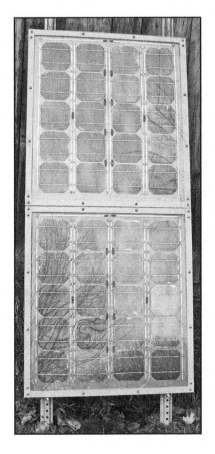

Above and left, 2.5 volt 20 amp panels; below, 2.5 volt 3.6 amp panel.

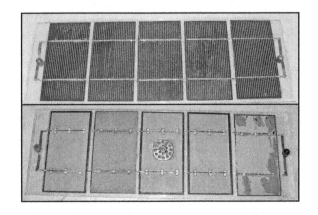

Solar panel components

Plexiglas sheet	Two 17½" X 10½" sheet and one 1½" X 1½" sheet. Local hardware store
Clear extruded acrylic rectangular bar	Purchase one ⅛"x ⅛"x 6 ft. From this bar, cut two 17½", two 10¼", one 15½". *McMaster Carr* 8728K11
Perforated base nut/binding nut	One zinc-plated steel, ¼"-20 thread size. *McMaster Carr* 98007A200
Plastic welder with Touch-N-Flow applicator	*Micro-Mark* 84131
Silicone sealant	Clear. Local hardware store.
Solar cells	Six cells 5"x 5" rated at 4 amps per cell. *Plastecs*.
Tab ribbon	16 pieces tab ribbon 10⅛". *Plastecs*
Bus ribbon	One piece 7¾", two pieces 2⅝". *Plastecs*
Flux-Pen	Rosin flux type RMA, Kester #186. Electronics store.
Solder	60/40. Electronics store.
Binding posts	Two. *Radio Shack* 274-661 or any other electronics store.

Parts numbers and suppliers have been provided for your convenience; however, suppliers may go out of business and parts numbers may change. All parts listed are available from multiple suppliers.

Tools needed

Plastic sheet cutter
Drill
Saw
Epoxy

Toothpicks for applicators
Soldering iron

Supercapacitor Holders

The individual cells that make up a supercapacitor power supply can be connected in series, parallel, or combinations of parallel and series. The power supply holder design depends on the dimensions of the supercapacitors and the way they will be connected. The connectors used must accommodate the type of leads on the individual supercapacitors.

Maxwell MC series supercapacitors come with either solid weldable terminals or threaded terminals. These can be connected by bus bars that you design or those that come with an integration

Left, threaded terminal; right, terminal with tab lead.

kit that Maxwell offers. The MC series can also be connected with ring terminals and braided wire or cable. I recommend the threaded terminals for experimental as well as most other applications. They are easily connected to the bus bar and ring connectors via nuts.

The Maxwell HC series supercapacitors have radial leads. These can be soldered or connected via Fahnstock clips. Fahnstock clips allow quick changes of configuration for experimenting.

If you use the Maxwell integration kit, the supercapacitors must have threaded terminals. The bus bars in the kit are for series connections. You can use the other components of the kit for parallel connections but you will have to fabricate a bus bar, or connect with ring terminals.

For C and D size supercapacitor cells such as the Maxwell BC series, any C or D size battery holder can be used for the cells. Most battery holders are molded plastic with spring grips and wire connections to the terminals. Battery hold-

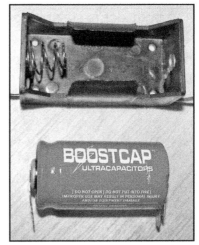

Single cell holder;

ers generally hold a single cell, or multiple cells in series connected configurations.

It is rare to find a parallel configured battery holder. If you need parallel connections for this size supercapacitor, you can connect multiple single battery holders, as shown below, by connecting all the positive wires together and all the negative wires together.

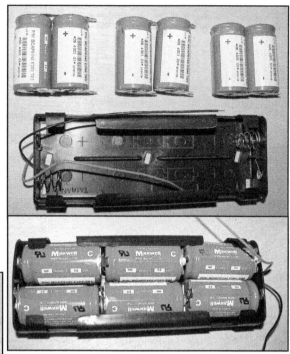

Fabricated parallel 6 cell holder, below..

Series 6 cell holder, above..

Parallel connected holder with copper bus bar for 140s

This is a parallel configured holder with copper foil or sheet for bus bars. The frame for the holder can be cut from PVC sheet or similar non-conductive material.

This holder is suitable for six Maxwell 140s connected in parallel with a total capacitance of 840 farads at 2.5 volts.

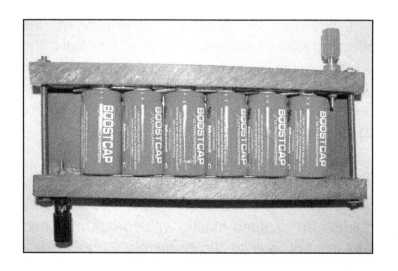

Parts list

Gray PVC (Type I) sheet	¼"x 12"x 12" *McMaster Carr* 8747K114
Gray PVC (Type I) sheet	⅛"x 12"x 12" *McMaster Carr* 8747K112
18-8 stainless steel threaded rod, 8-32 thread	6ft. length or 3ft. length. *McMaster Carr* 98920A009 (6ft.), 98847A009 (3ft.)
18-8 stainless steel hex serrated-flange nut, 8-32 screw size	Four needed, sold in packs of 50. *McMaster Carr* 93776A381 or hardware store
One pair of binding posts	Leads should be long enough to attach through ¼" PVC. *Radio Shack* 274-661
Copper foil, minimum .005" thickness	I used .005" copper foil, ½" wide, 100' roll. *McMaster Carr* 9053K14

Parts numbers and suppliers have been provided for your convenience; however, suppliers may go out of business and parts numbers may change. All parts listed are available from multiple suppliers.

The copper foil should be .005" or thicker, depending upon how much current will be drawn from or supplied to the supercaps. For most low current applications I use a .005, but you can also use copper bar. The holder dimensions here are for copper foil fit. I used two pieces of copper .005" foil, ½"x 7". If you use thick copper bar (for example ⅛" thick), the threaded rods must be long enough to accommodate the extra thickness of the bar.

Copper foil and sheet roof flashing can be purchased at hardware stores in several thicknesses. Craft stores often carry copper foil and sheet in smaller quantities, which may be less expensive if you have no other use for the foil. Copper bar can be purchased online from *McMaster Carr* or other hobby metal outlet.

Construction

◆ Cut two 8-32 threaded rods 3⅜" long.

◆ From the PVC sheet, cut two ¼"x 8"x 1" bars, and one ⅛"x 8"x 3" piece for the base plate.

◆ Drill two holes in each PVC bar to insert the 8-32 threaded rod. It is best to clamp the two bars together and drill the holes through both bars at the same time so that the holes are aligned to each other. If the holes are not aligned it will interfere with proper tightening and create problems.

◆ Cut two copper strips ½"x 7" from foil, sheet or bar stock.

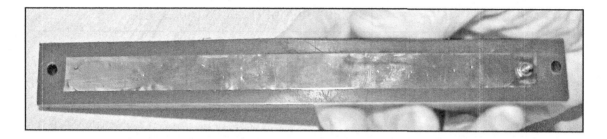

◆ Center and glue a foil strip to one side of each PVC bar. The foil strips must not be near the holes for the threaded rod because if the copper touches the rod it could create a dangerous short and cause a fire.

◆ Drill a hole in each bar for the binding posts. There must be enough space between binding posts and threaded rods to attach and spin the 8-32 nut freely into place on the threaded rod.

◆ Insert binding posts with nut and tighten so that good contact is made between nut and foil. Binding posts usually come in pairs, one red and one black. The red should always be used for positive input and output, and the black for negative input and output.

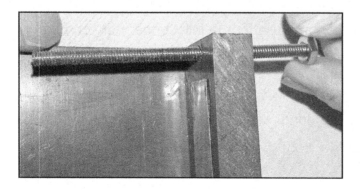

◆ Glue one PVC bar to the PVC base plate. The other bar will be free to move off the plate and is not glued to the base plate.

◆ Screw a 8-32 nut onto one end of each of the 8-32 rods and insert both through the bar attached to the base plate as shown at left.

- ◆ Insert the six supercapacitors.

- ◆ Slip the other PVC bar onto the 8-32 rods. Push until good contact is made with all the capacitors and attach the 8-32 nuts to the other side of the rod. Tighten firmly so that the capacitors are seated and cannot move in the holder.

- ◆ Once the supercapacitors are seated, attach input and output wires to the binding posts and the parallel supercapacitor bank is ready for use.

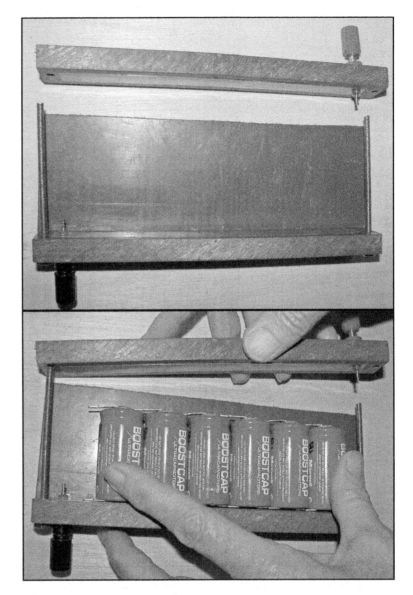

This holder can be made longer for more breathing room to isolate components such as binding posts and threaded rods, and otherwise modified according to your needs. Above all, the connections must be firm, not at all loose; and metallic components must not touch and create dangerous shorts. Although the voltage involved is not very high, the current capacity is very high, and can definitely melt metal and cause fires.

There are other versions of supercpacitor holders for parallel and series supplies with Maxwell 650s in the welding with supercaps (p.55) and the home PV system (p. 107) chapters.

Modified DC to DC Converter

Some applications require higher voltage than a 2.5 volt supercapacitor or bank provide, and a DC to DC converter can be used to provide the correct voltage.

Most DC to DC converters are used with vehicle plug attachments to provide 12 volts or less from a 12 volt automotive battery. These automotive converters can be modified with the simple addition of a boost converter wired into the fuse holder, or before the converter. The boost converter ups the voltage before it reaches the buck converter, then reduces the voltage in the buck converter (Recoton) to supply the voltage you will need.

There are many makes and models of portable converters on the market. We have chosen the *Recoton* to use as

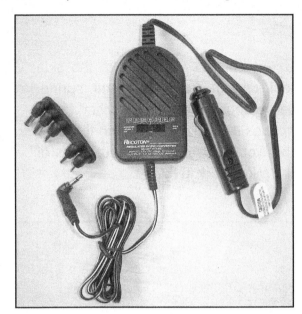

Recoton

an example of how to modify such a circuit. This particular Recoton model can supply up to 1 amp of current (maximum).

The *Micro Puck* boost converters used have a limit of 500 milliamps. This particular combination can power circuits of up to 500 milliamps due to the limitation of the boost converter. That being said, there are boost and buck converters on the market, or that you can build, that will allow much greater current use, if needed.

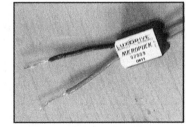

*Micro Puck
boost converter*

To modify the Recoton, open the fuse holder so that you can insert and connect the Micro Puck boost converter into the assembly.

When you have opened the fuse holder, note the wire that is attached to the coil spring in the center of the fuse holder. Mark this wire with a piece of tape. This is the positive input wire to the Recoton body.

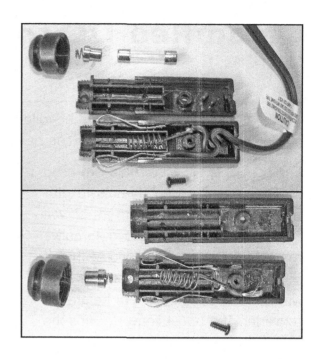

Cut the wires attached to the side leaf spring and center coil spring. Leave about ½" of wire attached to the springs. You will need to strip these so that you can solder the input wires of the Micro Puck to them.

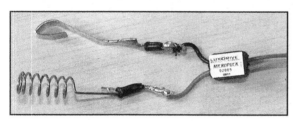

The Micro Puck input wires are red (positive) and black (negative). The black negative lead is connected to one of the side leaf springs. The red positive lead is connected to the center coil spring. Cut the red and black input wire to the Micro Puck to a short length so everything can fit into the fuse holder.

Strip these wires. Solder the red wire to the coil spring wire and the black wire to the side leaf spring wire.

Make sure the components are seated properly, reassemble the fuse holder and insert the fuse.

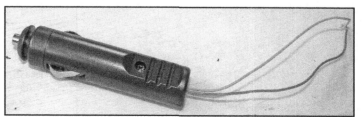

Strip the wires leading to the body of the Recoton and slip a piece of shrink tube over each of these two wires. This will insulate the two exposed wires from each other after they are soldered together. Over this, slip a larger diameter shrink tube to shrink over the other two for a more firm enclosure.

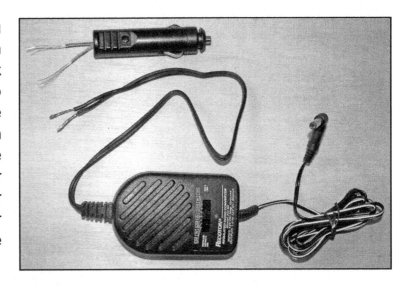

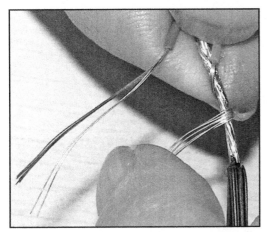

The Micro Puck output wires are orange (positive) and green (negative).

Solder the orange wire to the Recoton positive wire that you marked earlier. Solder the green output wire to the other (negative) wire from the Recoton.

Slip the shrink tube over the solder connections on each wire and heat. Slip the larger diameter tube over both wires and apply heat.

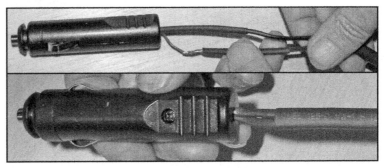

With this modification, a single 2.5 volt supercapacitor or 2.5 volt parallel wired supercapacitor bank can provide a variety of voltages up to 12 volts.

If you do not wish to insert the Micro Puck into the fuse holder assembly, you can simply wire it inline after the fuse holder and before the Recoton buck circuit.

Universal Solar Supercapacitor Power Supply

As an experimenter, I need a power supply that is flexible and portable. Sometimes I need AC and sometimes I need DC electricity. Sometimes I need high voltage, sometimes I need a specific low voltage. Sometimes I need the equipment in the laboratory and sometimes I need it in the field.

Of course everything has its limitations, but I have found that a dual 2.5 volt/15 volt solar supercapacitor supply with a few attachments is quite useful for a variety of situations and applications. The supply is simple in that it

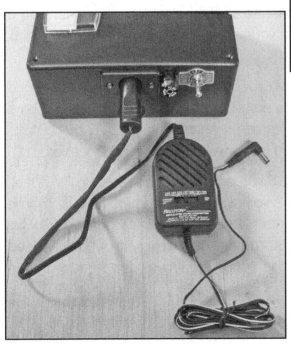

Above, universal power supply with modified Recoton.

contains one parallel connected supercapacitor bank, and one series connected supercapacitor bank. The 15 volt bank is charged by a 12 volt system PV panel and the 2.5 volt bank is powered by a 2.5 volt system panel. The power supply can be used while connected to the panels, or disconnected from the panels when the supercapacitors are charged up.

The 2.5 volt bank at left can be used with a modified DC to DC converter to provide up to 12 volts DC.

The 15 volt bank can power a regular DC to DC converter for voltages up to 12 volts and can also provide 120 volts AC by using a portable, very lightweight plug in AC inverter, at right.

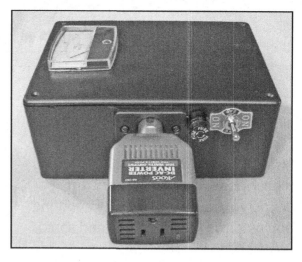
Universal power supply with AC inverter.

You may wonder, why bother with the supercapacitors for strictly daylight applications when you can just use a solar panel. The answer is that solar energy is intermittent, that is, clouds can temporarily block the sun. When this happens, power coming from the PV panels diminishes rapidly. The supercapacitor provides ride-through capability to keep the application powered until the sun comes out from behind that cloud. This is of great importance for vital communications and many other applications. Of course, the power supply can also provide night time power for applications such as LED lighting, etc. for short periods of time.

Building the power supply is fairly easy. You will need an enclosure that will easily fit the components. If you are not accustomed to stuffing electrical components in a box, the main thing to remember is to give yourself a lot more room than you think you need.

You can arrange components any way you see fit and a variety of connectors and components other than those mentioned in the parts list can be used. Terminals and bare wires must be insulated so that they do not come in contact and cause a short. Make sure that all the components, wiring, are rated for the voltage and current needed.

This particular supply can use either a 2.5 volt and or a 12 volt system PV panel. If you are using a commercial panel, the 12 volt system panel must have a charge controller to regulate and limit the voltage into the series connected bank to 15 volts. If you build your own 12 volt system panel you can design it to provide what you need more precisely without using a controller. You can use extra diodes to add voltage drop if necessary.

This design has two input jacks that connect the panels to the two supercapacitor banks. (See top of next page) Other types of input connectors such as binding posts can be used instead.

Above, the power supply has two input jacks.

The cells in one supercapacitor bank are connected in series and provide 15 volts, and the cells in the other bank are connected in parallel and provide 2.5 volts.

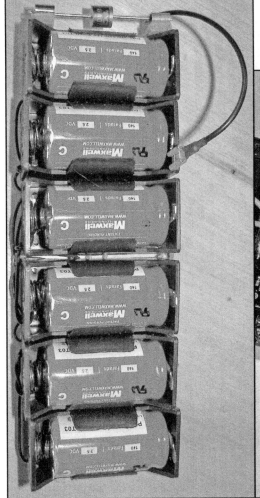

The 2.5v bank uses six 1 cell C size battery holders for the supercapacitors. The holders are connected to each other in parallel.

Below and left, 2.5 volt parallel supercap bank

The battery holders are attached to a Plexiglas sheet cut to form a base for the holders. Velcro strips are attached to the bottom of each cell holder and attached to the Plexiglas base.

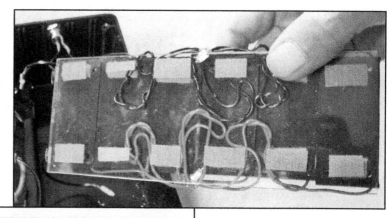

All of the black negative wires on the holders are connected and soldered together and all the red positive wires are connected and soldered together. The space between the bottoms of the C holders and the Plexiglas provides a place to tuck the wires.

The 15 volt bank uses one 6 cell series holder.

15 volt series supercap bank with balancing resistors, below and top of next page

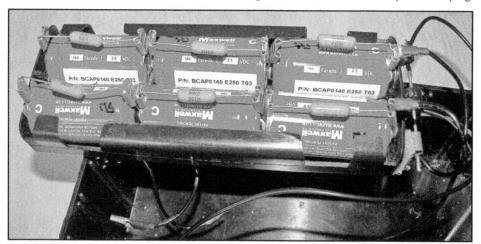

On one end of each holder I placed Fahnstock clips to hold the diodes. The clips are held onto the holders with Velcro tabs. You, can of course, place the diodes elsewhere and connect by soldering rather than using clips. If you use a charge controller on a 12 volt system panel you will probably not need a diode on the bank since most charge controllers contain a blocking diode.

Supercapacitor output is controlled by a DPDT toggle switch with a on-off-on position. This allows use of one bank or another or no output.

A fuse with a panel mount fuse holder is added to the circuit. The value of the fuse you use will depends on your needs.

Output is through an automotive type socket assembly and allows insertion of plugs for DC to DC converters and AC inverters.

Voltage is monitored by a 0-15 volt meter which is a useful reference to state of charge of the supercapacitors.

This design can be modified in many different ways. For instance, the output could include a house type outlet in the box to plug in AC powered devices with a wall plug. Once you understand how the box works, you can modify it for your own applications.

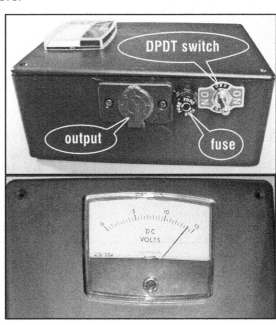

Double bank power supply schematic

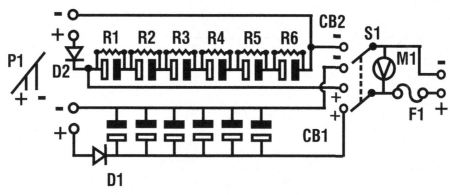

Legend

P1	Solar panel
R1 - R6	Resistors
D1, D2	Schottky or general rectifier diodes
CB1	Supercapacitor bank, 6 cells connected in parallel
CB2	Supercapacitor bank, 6 cells connected in series
S1	DPDT center off toggle switch
F1	Fuse 1 amp
M1	Meter 0-15 volt

Power supply parts & components

Project enclosure	8"x 6"x 3". *Radio Shack* 270-1809
Supercapacitors (Ultracapacitors)	6 or 12 BCAP140-250 Ultracapacitor: 140 Farad: 2.5 Volts. *Tecate Group*
Resistors	6, value according to need, for bypass resistance in series connected supercapacitors. *Allied Electronics*, others
Battery holder	Series connected six C cell size. *Thomas Distributing* MS12BH2636CW
Battery holders	6 one C cell size. *Jameco Electronics* 216346 or 216347

Plug and socket assembly	Model E2730. *The Prop Doc* 3000.0643
DC power jack	2, size M. *Radio Shack* 274-1563
DC power plug	2, size M. *Radio Shack* 274-1569
DPDT center off toggle switch	*Radio Shack* 275-1533
Panel meter	0-15 volt DC. *Radio Shack* 22-410
Panel mount fuse holder	*Radio Shack* 270-367
Fuse	1¼ x ¼", according to needs
Diodes	2, Schottky or general rectifier. *Galco Industrial Electronics*, *Allied Electronics*, *Backwoods Solar* and others.

Parts numbers and suppliers have been provided for your convenience; however, suppliers may go out of business and parts numbers may change. All parts listed are available from multiple suppliers.

Attachments and options

The Recoton AD 62-2 DC to DC converter can be modified with the addition of either a Micro Puck 500 milliamp DC to DC converter, or a Micro Puck DC to DC 350 milliamp boost converter, depending on your application's current needs.

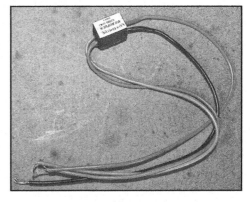

*Micro Puck DC-DC converter
500 milliamp*

Gang switches

Supercapacitors may be charged in parallel and discharged in series by including a gang switch in the power supply.

One of the advantages of using this type of switch is that you can build a dual 2.5 volt/15 volt power supply with only six capacitors instead of twelve. Also, the power supply can be charged with a 2.5 volt system PV panel, which is less cumbersome and expensive than a 12 volt system panel. Finally, if the supercapacitors are connected in parallel when charging, it is not necessary to have a balancing system of active circuitry or balancing resistors.

With a power supply set up with a gang switch, the system can be used while it is charging during daylight hours to provide either 2.5 volts, or if a modified Recoton is used, 12 volts. Another mode of operation would be to charge, disconnect, and then switch to series connection for discharge at 15 volts.

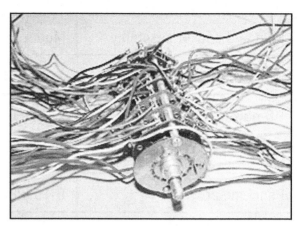

The Medusa, typical gang switch wired for one of the author's more complex supercapacitor projects.

Power supply with gang switch

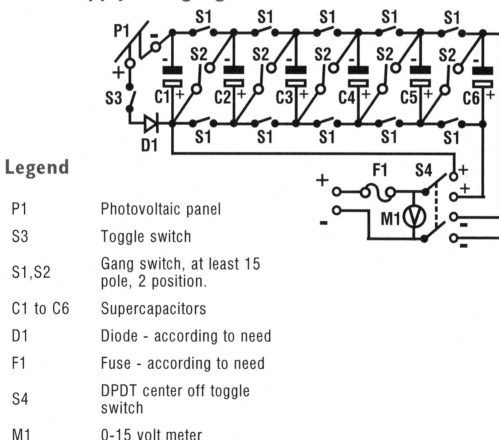

Legend

P1	Photovoltaic panel
S3	Toggle switch
S1, S2	Gang switch, at least 15 pole, 2 position.
C1 to C6	Supercapacitors
D1	Diode - according to need
F1	Fuse - according to need
S4	DPDT center off toggle switch
M1	0-15 volt meter

To charge the capacitors in a parallel configuration, all of the S1 switches are switched to the closed position and all the S2 switches are switched to the open position. This multiple switching is combined into a single operation by correctly wiring a 15 pole, 2 position gang switch to replace all of the S1 and S2 switches. When charging is completed, switch S3 is opened and S1 and S2 are reversed using the gang switch. The capacitors will then be connected in series and provide 15 volts. Switch S4 will determine the output, either 2.5 volts or 15 volts.

A wide variety of methods can be used. Switching, timing of switching, and charge and discharge times could all be automated with solar supercapacitor powered microcontrollers and other devices. The possibilities and configurations are endless.

Additions and options parts list

Recoton DC-DC buck converter	Model AD 62-2, regulated DC to DC converter 1.5 volt to 12 volt DC-positions at 1.5, 3, 4.5, 6, 7.5, 9, and 12 volt, selectable. *Herbach & Rademan Company* TMO3ADR4718. Can use any other
Micro Puck DC-DC boost converter	500 milliamp. *LED Supply* 02009-SHO
Micro Puck DC-DC boost converter	350 milliamp. *LED Supply* 02009A
Vehicle power adapter plug with banana jack binding post	*Radio Shack* 270-1521
Akoos AK100 DC-AC power inverter	100 watts/190 watts peak or other brand. *MCM Electronics* 58-13230. Many other suppliers are available for this product and others like it.
Rotary switch	15 pole 3 position rotary switch PA-2038 or similar (a 15 pole 2 position switch is all that is necessary) by Centralab *Sacramento Electronics Stores, Electroswitch Electronic Products, Galco Industrial Electronics*

Parts numbers and suppliers have been provided for your convenience; however, suppliers may go out of business and parts numbers may change. All parts listed are available from multiple suppliers.

Solar Supercapacitor Spot Welder

As an experimenter I often need to weld thin materials for precision scientific equipment such as electrolyzers, fuel cells, battery supplies and many other items. Such welding is not generally available in commercial shops at a reasonable price, so it was worthwhile for me to develop a solar supercapacitor spot welder.

Although the welder discussed here is for spot welding thin materials, this project can be scaled up for any type of electrical welding of any thickness using solar powered supercapacitors.

A basic spot welder consists of two copper electrodes and a power supply. One electrode is connected to the positive terminal of the supercapacitor bank and the other electrode is connected to the negative terminal of the supercapacitor bank. With such a welder a variety of complicated and durable shapes can be formed.

For most spot and seam welding a low voltage high current input is needed. This particular welding unit has six Maxwell 650 supercapacitors connected in parallel, which provide enough current to weld a variety of foils from .001" to .01".

To spot weld, the welder electrodes are applied to the metals to be joined. The metals to be joined have resistance to the current flowing between the electrodes. This creates heat, which melts and bonds the pieces together.

Solar panels for the welding unit

The welding unit uses either 2.5 volt or 2.7 volt system panels. A blocking diode is used to prevent reverse current flow. This can be any general rectifier diode or Schottky for low loss. The diodes must be rated for the voltage and current delivered by the panels.

The PV panel can be fabricated and configured to deliver more or less current depending on how fast you want the supercapacitors to charge. If you have a high duty cycle application, the PV panel should deliver 12 to 20 amperes of current or more. If your needs are minimal, a simple five or six cell panel that delivers around 4 amperes will suffice to recharge the supercapacitor bank. A panel that delivers less than 4 amperes can be used, but it will take longer to charge the supercapacitors.

The solar panel chapter (see page 23) has further information about how to build a panel to match the requirements of this application.

Electrodes for spot and seam welding

There is a wide variety of commercial electrodes available, or they can be made from copper wire, sheet, or pipe. Commercial electrodes are essentially high strength, electrically conductive copper alloys and are classed according to use by the RWMA (Resistance Welders Manufacturing Association). Commercial electrodes can be purchased from *McMaster Carr* and other suppliers.

There are several ways to apply the electrodes, depending on the shape of the work piece. For instance, to weld metal tabs to battery terminals, the tabbing material is placed on top of the terminal it is to be joined with, then both electrodes are applied close to each other on the surface of the tabbing material.

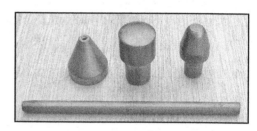

Commercial electrodes for welding

To bond two or more sheets of foil together, the electrodes are placed on opposite sides directly in line with each other and the current is applied to perform the weld.

Another technique is to use a conductive copper plate to rest two or more sheets on. A clamp electrode is attached to one pole of the supercapacitor bank and then clamped onto to the plate. The other electrode is a wire electrode connected to the opposite pole of the supercapacitor bank. The copper plate and the clamp electrode act as one electrode, and the current is conducted to the work piece sheets through the wire electrode as it is applied to the work piece.

Although a copper sheet is mentioned in this example for use with a clamp electrode, other forms, such as copper pipe for tubular shaped work pieces, for instance, can be used with clamp electrodes.

Care and use of electrodes

Electrodes oxidize readily upon use and have to be cleaned and surfaced after use. I brush the electrode across a Teflon scrubber pad, steel wool, or sandpaper to keep the surface free of oxidation after each application. Oxidation will definitely reduce the effectiveness of a weld, so it is important to have clean electrode surfaces.

Building the spot welder supercapacitor bank

This particular welding unit uses six Maxwell 650 MC series supercapacitors connected with aluminum bus bars in parallel

The aluminum bus bars were constructed for this application, and the plastic holders were purchased but there are many options for constructing supercapacitor banks. Any non-conductive plastic strips can be used as place holders for the supercapacitors, with a conductive aluminum strip for bus bar connectors. Do not use

anodized aluminum as the surface is basically non-conducting and will cause serious problems. Some alternatives to a bus bar for the parallel connections needed for this application are wire or wire braid, with ring connectors.

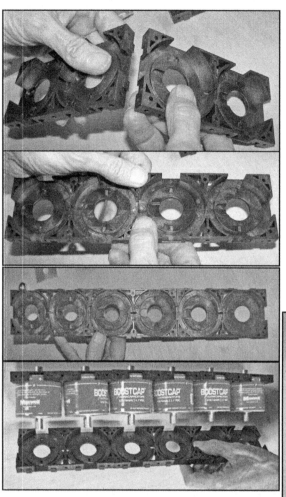

I used aluminum alloy 6061, .090" thick, 2"x 48" for bus bar stock. I cut two 13³/₄" lengths from this stock and drilled 6¹⁵/₃₂" holes centered on the stock along the length. Each hole from center to center was 2¹/₂"

The plastic holders and steel nuts and washers came from a Maxwell integration kit. I snapped the plastic holders together to form two long strips and then inserted the supercapacitors. I then placed the bus bars on each side of the supercapacitors and applied and tightened the nuts.

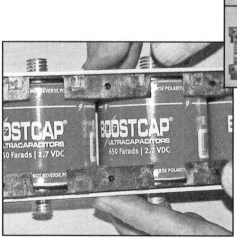

If you do not have parts to use from an integration kit you will have to find nuts and washers that will fit your supercapacitor terminals. See the parts list, page 63 for the sizes that will fit these particular supercapacitors.

The two welding units featured are easy to construct. One is used for tabbing batteries and the other is used for general purpose welding. A variety of wire sizes, switches, and designs can be used for building either unit. For either of the welding units, the ends of the wire electrodes should be rounded, slightly beveled, or otherwise shaped with a file or grinder for your particular application.

General purpose welding unit

For the general purpose welding unit, we used 10 gauge zip cord wire. Battery or welding cable could be used instead. The wire lengths for the welding unit should be as short as possible, but long enough to be comfortable and convenient to use.

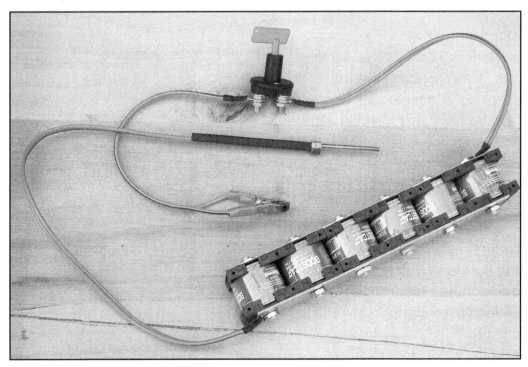

The zip cord wire is stripped at both ends. One end is attached to a ring terminal that fits the terminal of your supercapacitor bank, photo below.

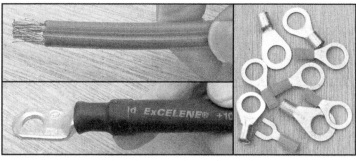

Zip cord, top left; battery cable, bottom left; ring and spade terminals, right.

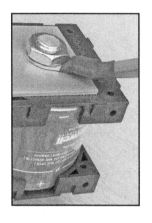

The other end of the zip cord will be attached to a solderless connector (shown at right). They can be found in most hardware stores in the electrical section. We used 8 gauge wire cut to length and slipped into the solderless connector along

Solderless connectors

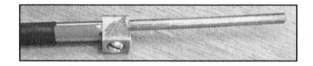

with the zip cord wire. The screw on the solderless connector is tightened to form a mechanical electrical bond between the 6 gauge wire electrode and the zip cord. Shrink tubing can be used to cover the wire at the handle area. For this particular unit I used a piece of 8 gauge wire that was $9^1/2$" long. 3" is bare and visible while the other $6^1/2$" forms and acts as a stiff handle for easy handling. Once the ends of the stripped 10 gauge wire and the 8 gauge wire are inserted into the connector, the connector screw is tightened firmly.

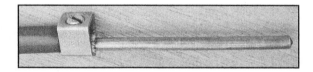

Above, cover the wire in the handle with shrink tube.
Below, electrode consisting of a battery clip and a conductive base plate

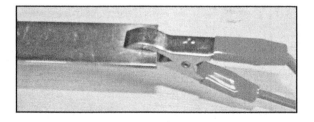

The other electrode for this unit was constructed from two pieces of 10 gauge zip cord. Each zip cord is stripped on both ends. A battery clip was attached to one of the zip cords. The clip will connect with a conductive base plate. Ring terminals that will fit the terminals of a high current switch and the supercapacitor bank terminal were added to the 3 other ends of the zip cord. The battery clip connects to the switch, which connects to the supercapacitor bank. The switch that we used for this particular unit is spring-loaded, so it can either be used as a momentary switch or a set on-off switch.

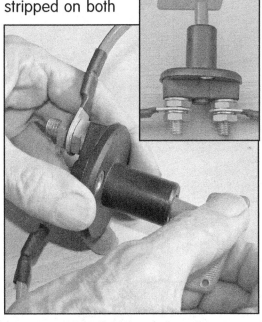

On/off and momentary high current switch

Momentary switch

You can also use just a momentary switch and set it up with a foot paddle for hands free operation.

Battery tab welder

The battery tab welder was constructed using 12 gauge zip cord, some old multi meter probe handles and smaller gauge electrode wire and two smaller solderless connectors, shrink tube and electrical tape.

No matter how the unit is constructed, it is critical that the two elec-

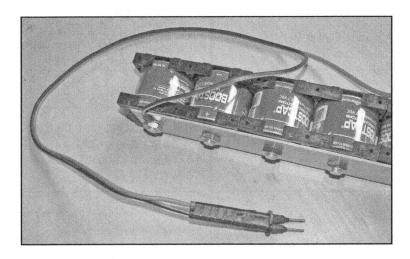

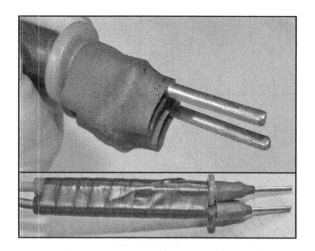

trodes are insulated from each by using shrink tube, electrical tape, or any other means to avoid a short. There are many ways to construct such a unit.

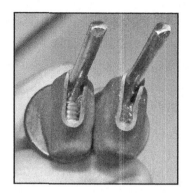

Welder schematic

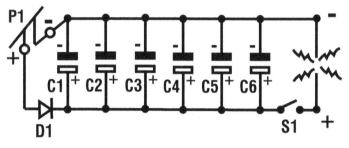

P1	Photovoltaic panel
D1	Schottky or general rectifier diode, rating according to need
S1	High current switch
C1 to C6	Supercapacitors

Parts and materials used for welder

Maxwell 650 farad 2.7 volt ultracapacitor	6, BCAP0650-P270 with threaded terminals. *Tecate Group*
Maxwell Power Burst Integration Kit,	For six cell modules. *Tecate Group*
Aluminum alloy 6061 stock	.090" thick, 2"x 48", for bus bar. *McMaster Carr* 89015K85
Battery master switch	On/off momentary. Maximum load for 10 seconds is 1000 amps at 12 volts, 500 amps at 24 volts. Continuous load rating 50 amps at 12 volts, continuous load rating 25 amps at 24 volts. *Wiring Products* 79997
Momentary push button switch	For optional foot switch. Heavy duty, rated 50 amps at 12 volts. *Wiring Products* 73475
Foot control switch	*Prop MD* 3001.5687

Parts and materials, continued

Battery clips	*Radio Shack* 270-343 or hardware or automotive supply.
Red and black zip cord	12, 10 and 8 gauge wire *The RF Connection*
Nickel alloy tabs, precut, for welding to battery packs	*Sunstone Engineering*
Nickel-iron foil	.002" through .01" thickness *McMaster Carr*, 8912K21-8912K34
Stainless steel sheet shim stock	.001" to .01" thickness, type 316 stainless steel *McMaster Carr* 2317K11- 2317K17
Monel/nickel and stainless steel mesh screen	*McMaster Carr*
Connectors	A variety of connectors will be needed, for the most part, crimp on connectors and terminal blocks. Terminal blocks should be high amp rating.
Clamp type battery connectors	Automotive parts or hardware store.
Bare copper wire	6, 8 or other gauge Mostly used for grounding. Hardware or electrical supply.
Copper solderless connectors	Hardware store, electrical section
If you do not have parts to use from an integration kit, these nuts and washers will fit the threaded terminals of the Maxwell BCAP0650-P270.	
Nuts	Metric 18-8 stainless steel thin hex nut, M12 size, 1.75mm pitch, 19mm width, 6mm height *McMaster Carr* 90710A130 or stainless steel McMaster Carr 93935A50
Wave washers	18-8 stainless steel single-wave washer M12 screw size, 24mm OD, 1.2mm thick *McMaster Carr* 92168A113
Flat washers	Any washer for M12 screw size

Parts numbers and suppliers have been provided for your convenience; however, suppliers may go out of business and parts numbers may change. All parts listed are available from multiple suppliers.

Suitable materials for spot and seam welding.

◆ Mild or low carbon steels such as austenitic in the 300 series and ferritic are easy to weld because their resistance is higher, so the current requirements are lower.

◆ Martensitic steels are very hard and produce brittle welds, so they are not good candidates for this process, although the tendency to brittleness can be reduced by tempering.

◆ Nickel iron alloy and nickel are good candidates for spot and seam welding, as are stainless steel and Monel mesh.

◆ Aluminum can be spot welded but requires high current due to its conductivity.

Some dissimilar materials such as copper and steel can be spot welded but not readily due to differences in conductivity, thermal, and melt properties.

Welding technique

Welding is an art as much as it is a science. With this simple welder, you have to gauge the timing of the application of current to the work piece as it is a hand mechanical action rather than a matter of using a timer as is found on more expensive commercial units. For most applications, a quick and firm touch is all that is necessary.

Generally, spot welding occurs quickly. If you leave the electrodes in operation, that is, pass the current for too long, you will burn a hole in the work piece. Thinner materials weld very quickly. Combinations of thicker and thinner materials vary according to how many layers are involved and what the total thickness is. If you have to leave the electrodes in place because work pieces aren't bonding, you are probably not applying enough current for the job.

Any unit you build will have its own limitations and you will have to experiment with different thicknesses and materials to see how it performs. Also, the charge depletion in the supercapacitor bank will affect welding capability. Note how many of a certain type of weld you can perform before you have to recharge the bank. If the bank is being charged while in use, it may still need to rest and build up a charge if the duty cycle or current needs are high. The surface area of the spot weld depends on the shape of the electrode used. Seam welds are easy to do and simply require the movement of one electrode over the area to be seamed at a pace that will not be too slow so as to burn a hole in the piece, or too fast wherein a weld does not occur. A little practice will indicate what is necessary. Welding fine metal mesh, whether Monel or stainless steel, will require a few practice runs before engaging with the actual piece as the mesh is easily burned through and generally needs a quicker touch.

Safety

The solar supercapacitor welder operates at a very low voltage, so it is not a shock hazard. It is, however, a fire and burn hazard.

- ◆ Never weld around flammable liquids, gases or any materials that can be ignited.
- ◆ Before welding, remove all jewelry. Metal on the body can produce extremely severe burns if it should become part of the circuit. Take this precaution seriously.
- ◆ Always wear safety glasses or face shield and other appropriate clothing when welding.
- ◆ Always include an on/off or momentary switch in the circuit.

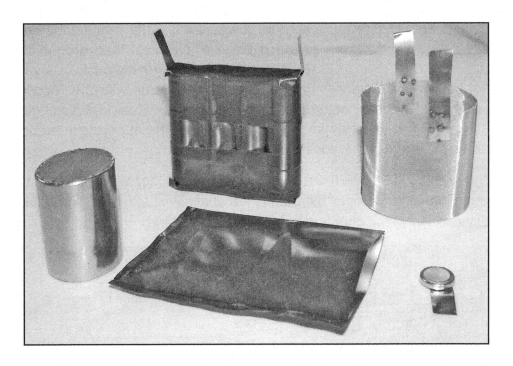

Welding Examples

There are many ways to approach spot and seam welding depending on the nature of the materials used and the shape desired. I have included several examples so that you can get some idea of what you can do with a simple solar supercapacitor spot and seam welding unit.

Simple rechargeable battery packs

Making battery packs can be very useful as you can replace the battery packs in a myriad of electronic devices and make specialty packs for electronics projects. All that is needed are rechargeable batteries; tabbing material such as nickel, nickel alloy, or stainless steel; good electrical tape, and a supercapacitor welder.

For higher voltage from the batteries, connect the batteries in series, as in the diagram at right.

Series connections

Parallel connections

For more current, if the voltage of one battery will suffice, you can connect the batteries in parallel as in the diagram at left.

Of course, you can also connect the batteries in combinations of parallel and series to meet other specific power needs. After aligning the batteries, wrap a piece of electrical tape around them to hold them firmly and form the pack. Depending on the type of battery, the distances from electrode to electrode will vary. Measure the distance and cut the tabbing accordingly. Foil of .003" to .005" thickness can generally be used, however this depends on the current draw of the application the batteries will power. High current draw requires thicker foil.

After the tabs are cut, place them across the terminals they will be welded to. Hold the tabbing material to the top of the battery terminal either with your gloved fingers or a piece of wood. Apply the double electrode firmly but quickly on the tabbing material,

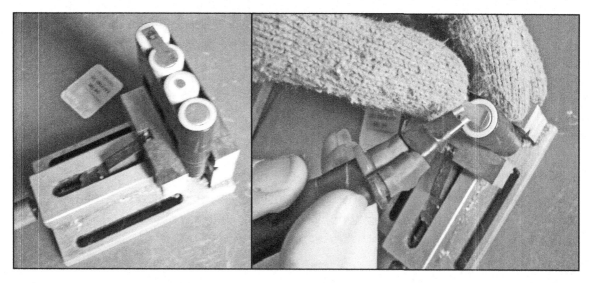

which should be in firm contact with the terminal it is being welded to. Basically a firm tap with the electrodes is all that is needed. Be sure that you do not hold the electrodes to the tabs and battery terminals too long as the battery can explode. Always wear appropriate safety clothing and a protective face mask when welding battery terminals.

When the welds are complete, wrap the pack neatly with electrical tape so that the only exposed surfaces are the tabs from the negative and positive terminal of the pack.

*Button battery
with tab weld*

Welding Mesh Cylinders

To fabricate small diameter mesh cylinders, I use a copper bar as one of the electrodes. In the example shown here, I use a $\frac{1}{8}$"x 1" copper bar. Copper pipe can be used instead.

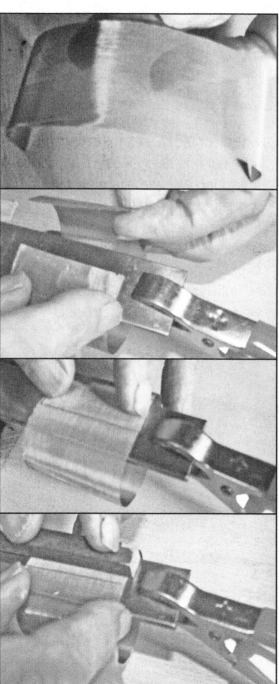

Attach a clip electrode to the bar as near to the piece being welded as possible, but not touching the piece itself.

Cut the mesh to the length and width needed. Place the piece around the bar and overlap the ends that will be welded together. To make sure that the overlapped ends do not move, use a flat wooden stick to hold them firmly against each other while you perform the weld.

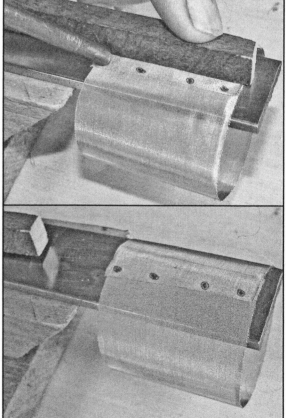

When the piece is positioned, apply the wire electrode at the spots where you want the welds. Space the weld spots as needed for your application.

Welding foil tabs and thicker metal components to mesh

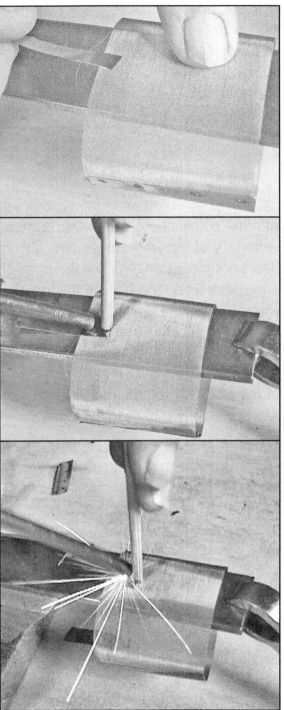

Some of my projects have required welding foil to mesh for electrolyzers, fuel cells, and a variety of scientific instruments. The process is the same as welding foil to foil or mesh to mesh. Lay the foil tab on the surface of the screen, hold it in position with a flat tipped wooden stick and apply the wire electrode to the spots you need to weld. Be sure there is good, firm contact between the mesh and foil to make a good weld.

The timing of the contact between the wire electrode and the work piece will vary according to the thickness and type of materials. In any case it will be short. Do not apply the current for too long or you will burn holes in your work pieces.

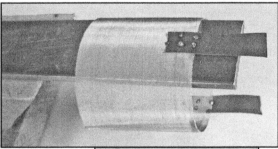

Filter screens

Occasionally I need to add filter screens to pipes and nipples, or add a screen to a thicker metal component for other purposes.

Use a flat copper plate for one of the electrodes and lay the thicker piece (such as a nipple) on the electrode. Cut the mesh to size, lay it on the surface it will be welded to and apply the wire electrode on the surfaces to be welded. Welding thin mesh to thicker metal pieces can be tricky because it is easy to burn through the mesh without making a weld. Practice on scrap pieces to get a feel for it.

Seam Welding

Rather than welding a spot, you may want make a continuous seam. This is easily accomplished by steadily moving across the surface of the work pieces with a wire electrode. Use a flat plate electrode to lay the work pieces on. Practice the pace of movement as you draw the electrode across the work piece. If you move too fast, the seam will not form; if you move too slowly, you will burn holes in the work pieces.

I sometimes make stainless steel bags to hold metal components for heat treating or other scientific experiments, or storage.

To make a bag, choose the thickness of foil you wish to use, and cut it to size (including margins for the seam along the edges) with a pair of sharp scissors.

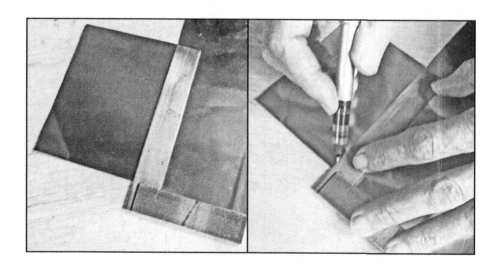

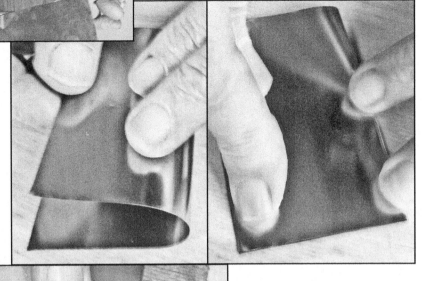

You can either run the seam right at the edge or inside a little bit along each edge. With a little practice you can see what works best for you.

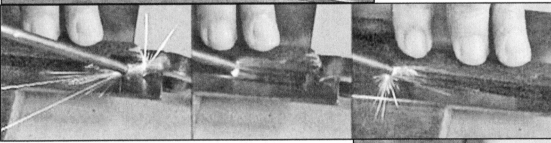

If the bags are to be used for long term storage or use in the field, the contents can be hermetically sealed inside the bag with a final weld.

Solar Supercapacitor Modification for Biomass Camping Stove

Over the years I have been interested in using the hydrogen rich gases that are wasted in the biomass burning process to improve portable outdoor cooking stove performance. I had read many articles about how to construct a variety of camp stoves that use this gas. Gasifiers, as these stoves are called, burn biomass more efficiently and minimize some pollutants.

I like outdoor cooking, and was interested in a fire that would come up to cooking heat fast, be consistent during a burn, and be something that I could really cook on without buying fuel. Forced draft gasifier stoves provide quick even heat, which is usually only supplied by stoves that burn fuel that must be purchased and carried.

Gasifiers can use any kind of biomass for fuel: pine cones, needles, twigs, leaves, grass, bark, charcoal, etc.

The designs that I had seen relied on batteries to drive the fan needed for the draft. I immediately thought that this type of stove would be a good candidate for a solar supercapacitor power supply. While researching biomass stoves to modify, I found a stove designed by Dr. Tom Reed and being offered by the *Biomass Energy Foundation*. Prior to the discovery of this stove offering, I had read *Superficial Velocity: The Key To Downdraft Gasification* by Dr. Reed and was impressed with his work. I realized that his design would be one darned good stove to work with. It was perfect for a make over into a more eco-friendly stove. The specifications looked very good.

The stainless steel stove comes in two sizes: XL (9"x 6.5", and 2 lbs. 3 oz), and LE (6.25" x 5.1" and 1 lb. 7 oz.). Both stoves require two AA batteries to operate the fan. The battery life estimate is 10 hours for the LE and 8 hours for the XL. The LE has a stated draw of 150mA at 3V on high, and 140mA at 3V on low. The XL has a stated draw of 180mA at 3V on high, and 170mA at 3V on low. The burn time for the LE is estimated at 10 to 12 minutes for a single charge using twigs and 40 to 50 minutes for a single charge using sawdust pellets, with an output of 10,000 BTU/hour on high and 5,000 BTU/hour on low.

The burn time for the XL is estimated at 20-24 minutes for single charge of twigs, and 70-84 minutes for single charge of sawdust pellets, with an output of 12,000 BTU/hour on high and 8,000 BTU/hour on low.

Reed's stove can be run continuously by adding fuel in small increments during use, so you can cook all day on it if you want to, and the fuel is literally right at your feet and free.

The only problem from an ecological and economic point of view is that it runs on batteries, which are expensive and quickly end up in the landfill. Now, I think batteries are great devices, but I try to eliminate their use as much as possible. It makes sense to use a solar supercapacitor system in this case. After all, why not plug into the sun if you can? Battery usage can be relegated to bad weather and emergency situations.

A solar supercapacitor biomass gasifier stove system will last the lifetime of your outdoor cooking career, provide home stove top performance, save you money and is an excellent application to promote greener technology.

While researching stoves to modify, I found others that would also make good conversion candidates, depending on your outdoor cooking needs. For instance, there is a forced draft stove offered by *ZZ Manufacturing* which they call a Sierra stove. It is made of titanium and only weighs 10 oz. This would be a perfect choice for the backpacker who has serious weight restrictions. I did not choose the Sierra stove because it did not fit my profile of use. I live in a rural area and mostly cook outdoors right outside of my house, or a short hike away. I also like to use heavy pots such as pressure cookers. Reed's stove has a robustness that served my purposes and can handle

heavier household pots and pans that are not appropriate for backpacking. If you are a backpacker with weight restrictions to consider, the Sierra stove could be modified for supercapacitors in the same way.

For expeditions, I would classify Reed's stove as a very stable, high performance, lightweight, small base camp stove for those areas where biomass is available for burning. In larger base camps it serves very well as a sidecar, high performance heating and cooking unit. I would even venture to suggest that packers and trekkers consider the smaller version of this stove as it is stable, easy to use, and performs very well. It might be worth the extra weight.

Components

The basic components of the power supply for the stove consists of a small solar panel, supercapacitor, a fuse, and a diode.

You will also need a project enclosure, Fahnstock clips (if you wish to use them), connecting wire, several power plugs and jacks or binding posts, and a terminal block for connections (optional).

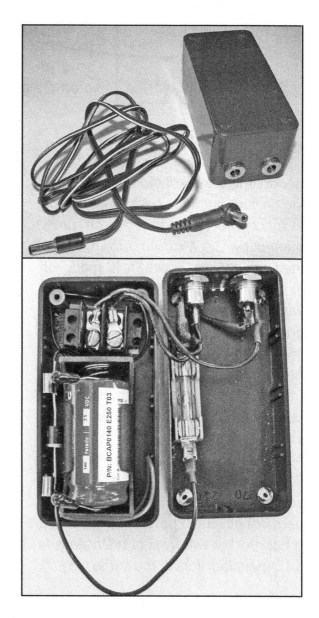

Fahnstock clips are available from *Ocean State Electronics* 2090, or other electronics store. I use the Fahnstock clips to hold the diode, but other types could be used. I prefer them because they permit me to change diodes easily for experimental purposes. Diodes come in different package designs, so some are not suitable for insertion into Fahnstock clips and would have to be soldered. Terminal blocks are available from any electronics store such as *Radio Shack*, or *Ocean State Electronics* (TB47302 - the two terminal block used in this example). Of course, if you wish, you can solder the connections and not use these connectors.

To connect the power supply output on the box to the stove, I used a *Radio Shack* 2741563 jack, and 274-1569 plug, and a pair of stranded 20 gauge stranded wires with a plug on each end.

To attach the solar panel to the power supply, I used a length of paired wires with a plug on one end to connect to the input on the power supply box. It is important to maintain the same polarities on the plug tips and jacks. I usually make the plug tip positive and thus the jack inside tip positive, and the outsides negative. Make sure the polarities match since a mistake will cause problems to say the least.

If you are new to soldering, you might find it easier to use a different type of power plug or binding posts. The plugs and jacks can be tricky to solder, depending on the type. If you do use the previously mentioned plugs and jacks, slip shrink tube over the wire, then slip it over the soldered connection, and shrink it. This will ensure isolation from other metallic components and wires and protect against shorts.

Be sure that all connections are solid, tight and insulated. When shorted out supercapacitors can melt metal, cause fires and at the very least will destroy your components.

For this project I used 20 gauge stranded wire. Stranded wire has more flexibility for bending than solid wire.

Supercapacitor configurations

The simplest configuration for this project is one supercapacitor. Although I used a 140 farad supercapacitor in the example, I suggest using a 2.7 volt 150 farad capacitor for better results since the fan was designed to be run off a 3 volt power source. The 2.5 volt capacitor works fine for my purposes, but the 2.7 will give a bit more oomph.

There are two jacks on the Reed stove, one for high and one for low. The low jack has a 100 ohm ¼ watt resistor connected to it, to reduce the current to the fan, which results in lower rpm. When using this stove with a supercapacitor, always use the high jack. With a 2.5 volt supercapacitor plugged into the high jack, you get the equivalent of plugging the battery supply that comes with the stove into the low jack.

To get the same results from the stove as using the batteries plugged into the high jack, you need to supply 3 volts or more. This can be done with two supercaps connected in series, and a rheostat or potentiometer to control the current and thus the fan speed (rpm). I used a small 3 watt 25 ohm rheostat. A 3 watt, 25 ohm rheostat such as *Radio Shack* 271-0265, or a 5 watt, 100 ohm potentiometer such as *Jameco* 140514 will suffice. Anything between 25 ohms to 100 ohms should be fine. The higher resistance of 100 ohms will give a finer control over the heat output. Be careful not to have the wiper blade W situated next to terminal A when you plug in the stove, as that will allow too much current to enter the fan since you are using a higher voltage. This could burn out the motor or create a blowtorch rather than a cooking stove fire.

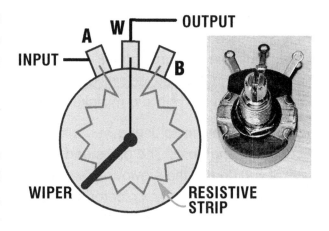

When starting the stove, the wiper blade should be near terminal B. Then, turn the wiper blade to the appropriate setting towards terminal A. Adjust the knob toward terminal A occasionally as the capacitor discharges to maintain the air flow and heat setting you desire.There are a lot of other possibilities for designing a power supply for this device. For instance, to get a longer run time from series connected supercapacitors you could connect three parallel connected pairs in series.

Theory of operation

The solar panel (P1) charges the supercapacitor (C1) which is then discharged through the stove fan either with the panel connected or not. The

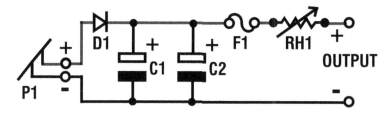

diode (D1) serves in a blocking function to prevent reverse discharge and current loss through the solar panel. The fuse (F1) provides circuit protection in case of a short in the circuit. If you use series connected supercapacitors, a rheostat (RH1) acts as a variable resistor that controls fan RPM and thus the heat available from the stove.

Above, power supply with two parallel capacitors and rheostat; below, single capacitor power supply.

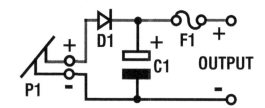

Schematic legend and suppliers

	P1	Solar panel, output considering diode voltage drop and capacitor voltage. Self fabricated according to need. See **section about solar panels.
	D1	Schottky diode or regular rectifier diode, rated for voltage and current of P1 solar panel output. I usually use 6 amp, 40 to 50 PIV which is more than I need but they are widely available such as Allied Electronics 502-0251 Schottky diode, rated at 45 volts, at 6 amps. Diodes are available in a variety of different package designs. Any electronics store
	F1	1/4"X 1" fuse and fuse holder, 1 or 2 amp slow blow fuse. A 1 amp fuse works fine for this project. Any electronics store
	C1	Supercapacitor, any HC, BC or MC series Maxwell supercapacitor, 140 farads or more. For a single capacitor I prefer a BCAP 0150-P270, though a BCAP 0140- E250 is fine. One supercapacitor can be used alone, or more can be used in parallel, series, or series parallel, depending on the stove and how it is to be used. Tecate Industries

Parts numbers and suppliers have been provided for your convenience; however, suppliers may go out of business and parts numbers may change. All parts listed are available from multiple suppliers.

Operating the stove

When using this stove with a supercapacitor, always use the high jack.

The stove can be directly powered by a solar panel through the supercapacitor during daylight hours. When not using the stove during daylight, disconnect the power supply from the stove and the supercap will charge for later use. A one time charge of the single capacitor power supply will give about 20 to 30 minutes of useful run time. The supercapacitor can also be charged at one location, then disconnected from the panel and carried to the cooking area for use; and, with the supercapacitor disconnected from the panel, other devices could be charged while cooking is occurring.

One of the advantages of this design is that you can run the stove on solar power even when the sun disappears behind some very dark clouds for a stretch of time. It reduces the intermittent failure that is common with strictly solar powered devices.

For really lousy weather when the rain just won't stop, supercapacitors can be charged with a small hand crank generator modified from commonly available flash-

lights. You could crank and charge while hiking, then when you are ready to rest and cook, the supercapacitors will be ready to go. You can, of course, also use the standard AA battery pack as backup. Although I use a rigid solar panel for this project, you could use flexible cells sewn into backpacks, tents, rolls, or jackets to charge the supercapacitors.

We really enjoy outdoor cooking with this solar-supercap modified stove.

Dr. Reed's stoves can be purchased at the *Biomass Energy Foundation.*

If you are a backpacker and want a lighter stove to experiment with, you can find the Sierra Titanium at *ZZ Manufacturing.*

To build and design your own stove, study the pages at the *Renewable Energy Policy Project.*

Solar Supercap Powered Wireless Rocket Ignition System

Despite wonderful advances in battery technology, batteries are not well suited for igniting rocket motors. Batteries are drained very quickly when used for ignition purposes and require frequent replacement even if rechargeable batteries are used. The cost of battery replacement adds up quickly.

Supercapacitors, on the other hand, have an almost limitless life and can be recharged very quickly. A supercapacitor has a lifetime of at least 100,000 cycles. This means that if you flew a rocket a day, your supercapacitor would last for 23 years. To put it another way, you could launch at least 100,000 rockets with one supercapacitor. This alone makes supercapacitors far superior to any present battery technology for rocket motor ignition.

If you combine the advantages of supercapacitors with photovoltaic charging you come up with a solid winner for powering rocket ignition. A solar supercapacitor rocket ignition system is environmentally friendly in that it reduces the waste of discarded batteries and uses an earth-friendly photon harvesting technique for charging. You never need to purchase batteries or pay for grid electricity to charge them, and you're not using power created by fossil fuel. It is also handy to be able to plug into the sun as an energy source while in the field.

Wireless ignition for rockets is another technology that adds convenience. It allows a freedom of movement that a cumbersome hard wire system cannot give. Once you have tried wireless ignition, you will be hooked.

Concept and components

Most rocketeers enjoy their hobby on sunny, or bright cloudy days. Given that, it makes sense to use the sun to provide the energy needed to launch rockets.

A solar supercapacitor system will work under a variety of light conditions. Of course, the more sunlight available, the quicker the supercapacitors will charge. That being said, there are always those days that start out sunny and then deteriorate to very dark cloud cover. For less than adequate light conditions you can integrate a battery or other type of backup energy source such as crank-fire into the system. For backup I prefer to use crank-fire as it is more eco-friendly and less costly in the long run than using batteries.

A basic solar supercapacitor system for rocket ignition includes a photovoltaic panel and a supercapacitor. The photovoltaic panel provides the energy to charge the supercapacitor. When enough energy is stored in the supercapacitor, a switch transfers that energy to the igniter of a rocket motor. The system can be hard wired, or configured as a wireless system depending on your preferences. The most simple solar supercapacitor ignition system would consist of:

- One 2.5 to 3 volt solar panel
- One blocking diode
- One high capacity supercapacitor
- Wiring
- Firing switch

The more complicated system detailed here consists of:

- One to three photovoltaic panels
- Several blocking diodes
- Three supercapacitors
- Transmitter
- Receiver
- Four DC to DC converters
- A meter to monitor charge

The system can be used with homemade igniters and Estes type igniters either as is, or modified. Once you understand how the circuit works, it is not difficult to modify it to suit your needs.

Transmitter schematic

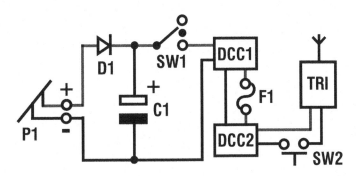

P1	Solar panel
C1	Supercapacitor
D1	Diode
SW1	Key switch 2SPST-NO, left: on, center: off key removal, right: on.
DCC1	Micro puck, 350 milliamp, DC to DC boost converter.
DCC2	Recoton DC to DC buck converter
F1	Fuse 1 amp
TR1	Transmitter/encoder module 433MHz with dip switch
SW2	Momentary push button switch,
TB1	Terminal block

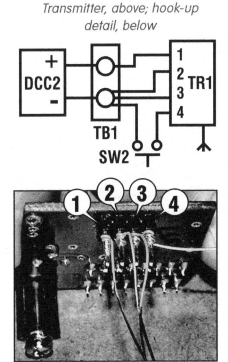

Transmitter, above; hook-up detail, below

Transmitter parts

	P1	Solar panel - self fabricated, see PV panels chapter**
	D1	Schottky diode or regular diode rated at least 6 to 10 amps at 40 to 50 PIV.
	C1	Maxwell 140F supercapacitor BCAP140-250 supercapacitor: 140 farad: 2.5 volts available from Tecate Industries. Can use other brands and values but should be at least 140 farads.
	SW1	Key switch 2SPST-NO, left: on, center: off key removal, right: on. McMaster Carr #728K18.
	DCC1	Micro Puck 02009A, 350 milliamp, DC to DC boost converter. LED Supply
	F1	Fuse 1 amp, ¼"x 1" with fuse holder. Any electronics store.
	SW2	Momentary push button switch, rated 3 amp at 12v AC, 1.5 amp at 250v AC. Radio Shack 275-609
	DCC2	Recoton DC to DC buck converter, model AD62-2, regulated DC to DC converter 1.5v to 12v, DC position at 1.5, 3, 4.5, 6, 7.5, 9, 12 volts, selectable. Herbach and Rademan TM03ADR4718 Can use any other brands.
	TR1	Transmitter/encoder module 433MHz with dip switch. Ramsey Electronics TXF433A . Can substitute with LINX, Ardino, Jaycar or others. Please note - with substitutions pin positions and voltage requirements may be different. Adjust accordingly

 TB1   Terminal block, two position, Radio Shack 274-656

Receiver schematic

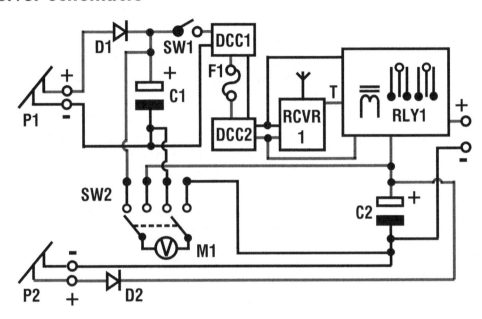

P1, P2	Solar panels	SW2	Toggle switch DPDT
C1, C2	Supercapacitors	RCVR1	Receiver decoder module
D1, D2	Diodes	RLY1	Sensitive relay DPDT
DCC1	Micro puck DC to DC boost converter.	M1	Panel meter
DCC2	Recoton DC to DC buck converter		
F1	Fuse 1 amp		
SW1	Toggle switch SPST		

Receiver hook-up details

DCC2	Recoton DC buck converter
RCVR1	Receiver decoder module
RLY	Sensitive relay DPDT
TB1	Terminal block

Receiver parts

	P1, P2	Solar panels, self fabricated, see chapter about PV panels
	RCVR1	Receiver decoder module 433 MHz receiver/decoder module with dip switch. Ramsey Electronics RXD 433A. Can substitute with LINX, Ardino, Jaycar or others. Please note - with substitutions pin positions and voltage requirements may be different. Adjust accordingly
	C1, C2	Six Maxwell 140F supercapacitors BCAP140-250 supercapacitor: 140 farad: 2.5 volts. Tecate Industries. Can use other brands and values but should be at least 140 farads.
	D1, D2	Schottky diode or regular diode rated at least 6 to 10 amps at 40 to 50 PIV.
	DCC1	Micro Puck DC to DC converter, 500 milliamp, 02009_SHO. LED Supply
	SW1	Toggle switch, SPST panel mount toggle switch, rated 20 amps at 12VDC. Radio Shack 275-601
	DCC2	Recoton DC to DC buck converter, model AD62-2, regulated DC to DC converter 1.5v to 12v, DC position at 1.5, 3, 4.5, 6, 7.5, 9, 12 volts selectable. Herbach and Rademan TM03ADR4718. Can use any other brands.
	SW2	Toggle switch, DPDT panel mount center off toggle switch Radio Shack 275-1533

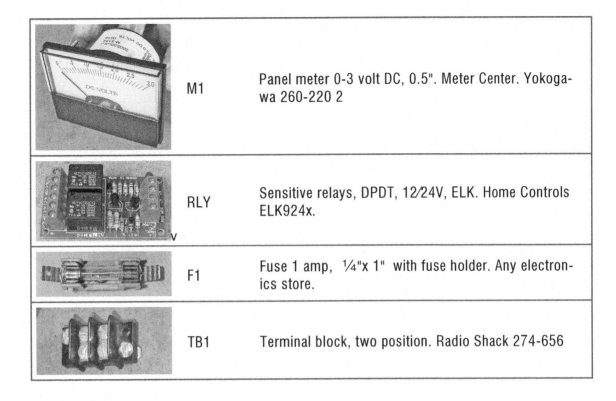

	M1	Panel meter 0-3 volt DC, 0.5". Meter Center. Yokogawa 260-220 2
	RLY	Sensitive relays, DPDT, 12/24V, ELK. Home Controls ELK924x.
	F1	Fuse 1 amp, ¼"x 1" with fuse holder. Any electronics store.
	TB1	Terminal block, two position. Radio Shack 274-656

Miscellaneous parts

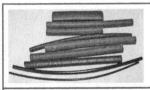

Shrink tube in a variety of sizes. Electronic stores such as Radio Shack 278-1627.

Wire wrap and tool. Electronic stores such as Radio Shack

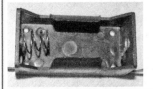

Battery holders for supercapacitors, Three C size. McMaster Carr 7712K411, or other types available from suppliers such as Jameco or other electronics supplier.

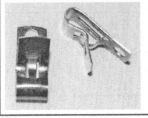

Fahnstock clips, alternative to soldering, optional. Electronics stores.

Miscellaneous parts, continued

Perf board, several pieces at least 6"X 8". Any electronics store such as Radio Shack 276-3396A.

Enclosure, transmitter. Serpac 17-5-13 electronics enclosure, Mouser Electronics 635-17-5-B. Any other suitable enclosure can be used.

Enclosure, receiver 7"x 5"x 3". Radio Shack 270-1807

Banana plugs for receiver output. Radio Shack 274-007 (accepts up to 12 gauge wire).
Dual binding post banana jack for receiver output. Radio Shack 274-0718

3 DC power jacks for solar panel input. Jack is for 5.5mm OD x 2.1mm ID tip size (one of many tip options for the Recoton). Radio Shack 274-1563. You can use any other tip sizes and simply purchase a jack for the tip size you choose. Jacks require $7/16$" mounting hole.
3 DC power plugs for solar panel input, 5.5mm OD x 2.1mm ID tip size. If you re-purpose the output wires on the Recoton these are already attached. Radio Shack 274-1569

Antenna covers, optional. Consists of flexible straw, rubber grommet, heat shrink tubing.

Toothless alligator clips, copper. This is crimped on to the 14 gauge two conductor stranded wire for attachment to rocket igniter. McMaster Carr 7236K51. Any other type of alligator clip can be used.

Insulated standoffs, such as Radio Shack 276-1381 1.

	Terminal block, 6 position, such as Radio Shack 274-659.
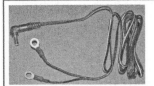	Re-purposed parts from Recoton.
	Optional quick connect terminals (female) to fit 0.250 and 0.187 male terminals. Herbach and Rademan Q5020 and Q5021
	Crimp on spade and ring terminals, variety assortment. Radio Shack 64-407.
	Wire, 18 gauge, 20 gauge and 22 gauge stranded hookup wire. Any electronics store. Also about five to six feet of two conductor 14 gauge stranded wire for hookup from receiver to rocket igniter. Can also use zipcord wire.
	Antenna wire, 2 pieces, six to seven inches ea. of #22 stranded wire or other.

Parts numbers and suppliers have been provided for your convenience; however, suppliers may go out of business and parts numbers may change. All parts listed are available from multiple suppliers.

Tools needed

- Saw to cut perf board
- Electric drill
- Hole cutter
- Reamer
- Nibbling tool
- Wire wrap tool

- Wire cutter
- Wire stripper
- Crimping tool
- Glue
- Soldering iron

Theory of operation

The solar panel supplies energy to the supercapacitor for storage through a diode that blocks reverse current flow.

When the safety key switch is turned to the "on" position and the firing push button is depressed, the charged supercapacitor supplies current and voltage through the boost and buck converter to activate the transmitter.

After being activated with a safety "on" switch, the receiver relay receives a coded signal from the transmitter, and discharges a supercapacitor to the output leading to the rocket igniter. The rocket igniter ignites the propellant and fires the rocket.

Building the transmitter and receiver

Open up the Recoton body and mark positive input and positive output with a red marker on the circuit board.

To do this, open the fuse holder and note which wire comes from the side spring negative polarity to the Recoton, and which one is connected to the positive tip of the fuse holder. Trace these to the Recoton circuit

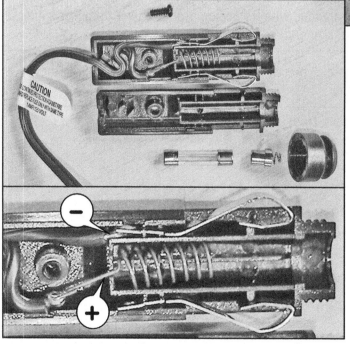

board and mark the positive lead or make a mark next to the positive lead on the circuit board. This will distinguish the positive lead from the negative lead visually when you hook up the wires. If these leads are not marked, it can be very confusing.

To determine the output polarities, either supply 12 volts to the input and then check the polarity of the output with a multi meter; or use the continuity tester on a multi meter. To test for continuity in order to determine polarity, set the meter to continuity and touch one lead of the multi meter to the marked positive input on the circuit board bottom where the wire is soldered, and touch the other lead to the opposite side of the board on the bottom solder points of one of the outputs. One of the leads will emit a tone which indicates continuity. Mark that output with a red marker on the circuit board next to the positive output wire. Do not test for continuity with power applied to the Recoton.

Cut the input and output wires at the point where the molded cord pull stop ends near the circuit board. This will leave a little bit of wire for connecting. Strip the input and output wires on the circuit board so that you can solder these wires to other wires that connect to components of the device.

Strip the ends of the output wires that were just cut (the wires that go to the plug heads, see photo) and attach terminals to the ends, see below. These terminals should fit the terminals on the

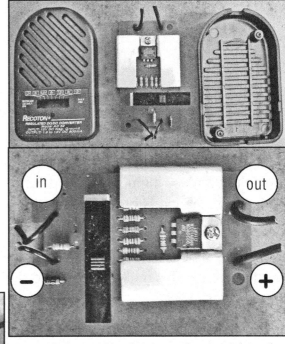

Mark the positive leads on the circuit board

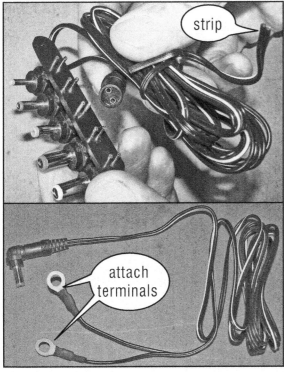

solar panel and will be the input connections from the solar panel.

One Recoton is needed for the transmitter and one for the receiver, so you might as well prepare both of them at the same time.

If you use the LED from the Recoton as an indicator on your project boxes, note that the positive lead of the LED is the one nearest to the resistor. Cut the LED leads halfway between the circuit board and the LED.

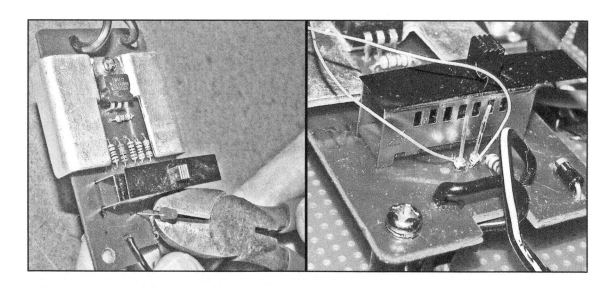

This should leave enough wire to wire wrap the leads sufficiently. Mark which severed lead on the LED is positive so that when you reattach the leads with wire wrap you do not confuse them. If the LED polarity is wrong, the LED will burn out and will not work. When the leads are cut, wire wrap a length of wire from the LED leads to the leads on the circuit board.

The wire should be long enough so that you can open the box and lay the top down beside the box. This makes it easier to make modifications without disconnecting or cutting wires.

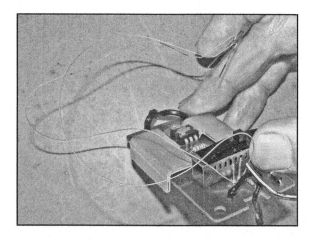

Do this for both the transmitter and receiver LED leads. They will be in different boxes with different dimensions, so allow plenty of wire length. I usually guesstimate and leave much more wire than necessary. Too long is generally better than too short. The wires can always be squished in. This same practice should also be followed for leads to panel meters and switches to perf board to ensure plenty of elbow room.

Remove the fuse from the lighter plug/fuse holder and put it aside to be inserted into a fuse holder in the box later.

Building the transmitter

The transmitter consists of about nine major components, an array of minor sub-components, a box and a perf board platform for the components.

The project box itself can be any type of box. If you have never constructed a project like this before or have limited experience in arranging components in tight spaces, use a box bigger than used here. A different shape is fine.

The perf board is cut with a saw to fit the box. Each box is different, with different molding patterns which protrude at

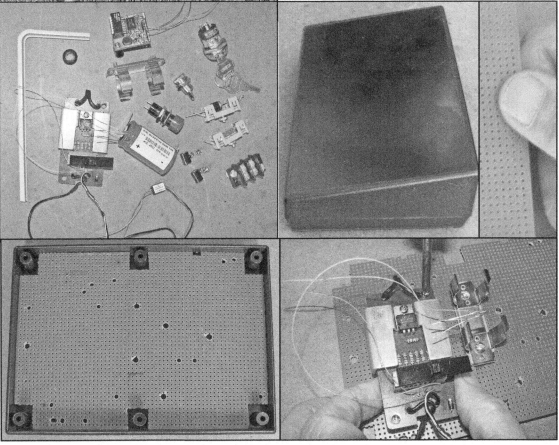

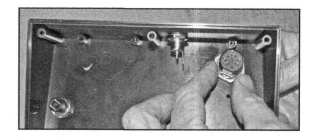

different points. Most have built in stand-offs where you can rest your board on the bottom of the box. Shape the board and drill holes to secure the board to the molded standoffs in the box.

Once the board is sized and configured, lay out the components on the board and work out their placement.Figure out the locations for the input and output jacks, holes for LED, key switch, push button switch, and antenna placement hole on the sides and outside cover of the box. The photos here show one way to lay out the components, but there are many ways to do this.

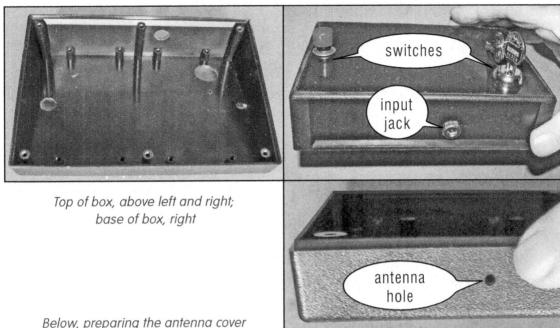

Top of box, above left and right; base of box, right

Below, preparing the antenna cover

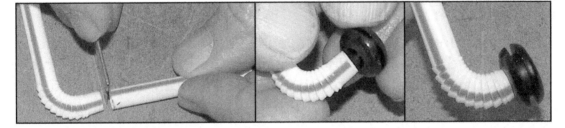

Transmitter circuit and component logic

- ◆ There is one input jack.

- ◆ The positive connection to the jack is connected to the diode to block and prevent reverse flow of current.

- ◆ The negative of the jack is directly connected to the negative terminal of the supercapacitor.

- ◆ The positive output from the diode is connected to the positive terminal of the supercapacitor.

- ◆ The positive terminal of the supercapacitor is connected to key arming Switch 1 which is also connected to the red positive input lead of the Micro Puck.

- ◆ The negative terminal of the supercapacitor is connected to the black negative lead of the Micro Puck.

Transmitter circuit and component logic, continued

- The orange positive output lead from the Micro Puck is connected to one side of the fuse.

- Another wire connects the other side of the fuse to the positive input terminal of the Recoton.

- The green negative lead from the Recoton is directly connected to the negative input terminal of the Recoton.

- The positive lead of the Recoton is connected as shown in the schematics and photos and the negative output lead is connected to push button firing Switch 2, which is also connected to the transmitter as shown in the schematics and photos.

- A simple 6" to 7" wire is soldered to the transmitter antenna connection

Charge the supercapacitors and you are ready to go.

Transmitter module

This Ramsey transmitter can be coded by the mechanical dip switches on the circuit board of the transmitter, or can be interfaced with a basic stamp or computer for coding. For that matter, this circuit can be totally stamp or computer controlled if you wish to design it that way. The coding provides unique addresses for transmission for safety. The transmitter sends a code that only your similarly coded receiver will pick up to initiate ignition. The transmitter range is about 300'.

This transmitter module comes with an instruction booklet which contains all the information needed to code with the dip switches, and includes hook up suggestions. Many other types of transmitter and receiver combinations could be used. This project is simply an introduction to the possibilities.

Connectors and connections

I used a variety of connectors, but any type of connector that suits the purpose is fine. I also used a terminal block for connections. This helps sometimes to keep things neat, but is not necessary. You can directly solder wire connections without using a terminal block. Similarly, I used Fahnstock clips for diode connections and holders, but you can just as easily solder these connections if you prefer. The female quick connect terminals for the supercapacitor connections could also be soldered directly with wire, or other types of connectors can be used depending on the terminal configuration of your particular supercapacitor. For instance, some supercapacitors have radial leads, so a battery holder type device would not be appropriate. You would need to solder the connections and design a holder that would fit your supercaps.

Building the receiver

If you do not mind waiting a little when charging, you can use one solar panel for charging both the capacitors in the receiver and the single capacitor in the transmitter. Or, you can set up two separate panels, one to charge both capacitors in the receiver at the same time and keep the charge ongoing during use; and the other panel to charge the transmitter.

I find that one small panel will do the job quickly enough for me. The capacitors will hold a charge for a number of launches before needing recharge. While the panels are charging, I simply do prep work on the rockets for the next series of launches.

Any type of project box can be used to house the receiver components as long as it is big enough to fit everything in. Cut and size the perf board for the box, figure out the component layout and drill holes to accept components, standoffs, terminal blocks etc.

Standoffs installed on perf board

Then, cut holes for antenna, meter, switches and power input and output terminals and LED. You can either follow the layout pattern in the photos or design your own configuration. Use

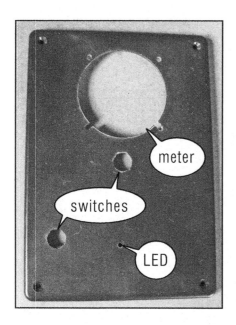

insulated wire throughout and do not let components touch each other in your layout. I use shrink tube to cover up the bare wire ends on connectors to provide electrical insulation.

As with the transmitter, I used a wide variety of connectors, Fahnstock clips, and terminal blocks. Each of these can be modified or changed to suit your needs.

Schottky diodes would be best, as with the transmitter, but any rectifier diode rated for the current and voltage will suffice if you take into account the voltage drop for charging, and modify the panels to increase their output to compensate for the loss.

The meter switch on the receiver box has three positions. One position is off, and the other two positions show the charge of capacitor one or two when you flip the switch to the left or right. This meter and switch combination is not necessary since it is possible to guesstimate when the capacitors are charged depending on sun conditions. I used a meter that had a scale of 3 volts as this is most appropriate, but a meter with less accuracy with a higher voltage range such as 5 to 10 volts could be used. The advantage of using a meter with a wider voltage range is that they are commonly available at a low price. A 3 volt meter is considered a specialty meter and is usually more expensive. You can save some money by making the leads available for a multi meter so that you can check the charge voltage of the capacitors via the multi meter.

The 3 volt meter used in the receiver

I did not put a meter on the transmitter; however, a meter could be added. It is not as critical as it is for the receiver, since the transmitter does not use as much energy as the receiver and relay.

Receiver component logic

The circuit/component logic of the receiver is simple.

Safety switch

- The receiver has two input jacks, P1 and P2.

- P2 supplies energy to charge supercapacitor C2 which in turn supplies energy to the rocket igniter.

- P1 supplies energy to charge supercapacitor C1 which powers the receiver and relay through the two DC to DC converters.

- When safety switch SW1 is closed, the receiver is powered up to receive signals from the transmitter.

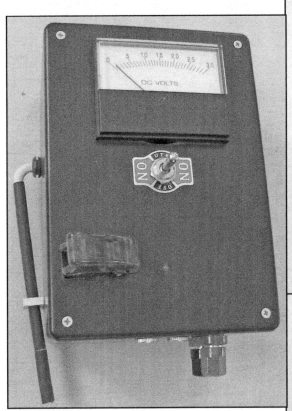

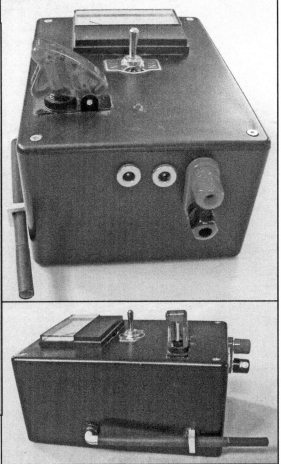

- ◆ When a coded signal is received, a trigger signal is sent to the relay which then discharges C2 into the igniter resistance wire.

- ◆ The resistance wire glows red and ignites the propellant in the rocket motor.

- ◆ When the signal from transmitter TR1 stops, the relay resumes its normally open position and the igniter is no longer energized.

Receiver circuit

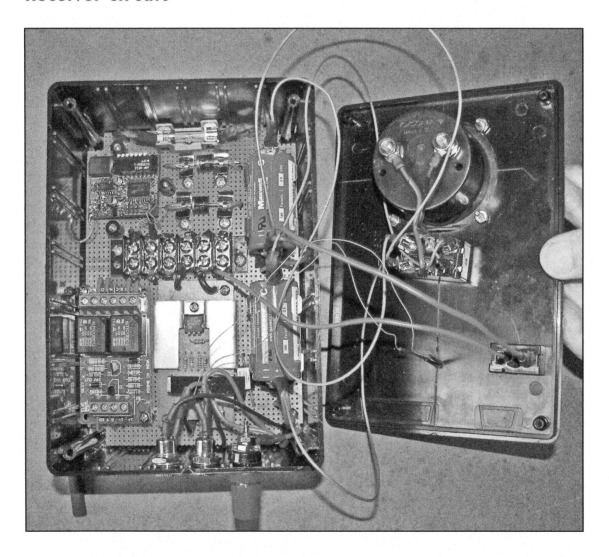

- ◆ The ELK 924 contains two relays that, for our purposes, are wired in parallel and thus can switch up to 12 amps. An instruction hook up sheet comes with each relay. There is also an instruction hook up manual that comes with the receiver.

- Two input jacks are needed for the receiver. One goes to a supercapacitor that powers the receiver and relay through the two DC to DC converters. The other jack charges the supercapacitor that supplies the energy for the igniter.

- Positive wire of each jack is connected to separate diodes to prevent reverse current flow.

- Positive output wire from diode D1 connected to P1 is connected to SW1, supercapacitor C1 and one of the terminals of SW2 which goes to 3 volt analog DC meter M1.

- Negative input from P1 goes to negative terminal of supercapacitor C1, one of the terminals on SW2, and to negative black terminal wire leading to the Micro Puck DCC1 boost convertor.

- Output (orange wire) of DCC1 goes to one side of the fuse.

- The other side of the fuse is connected to positive input of DCC2.

- Positive output from DCC2 goes to receiver input and to the relay power input terminals.

- Negative output from DCC2 goes to negative relay power input terminals.

- A wire is connected between trigger signal output of the receiver (referred to in the receiver manual as the DV or data valid output) to the positive trigger signal terminal on the relay.

- Positive output from P2 goes to diode D1 which is connected to the positive terminal of C2, to positive input of RLY1, and to one of the terminals of the 3 volt analog meter M1.

- Negative output from P2 goes to negative terminal of C2, to one of the terminals on M1, and to the black negative of the binding post terminal.

- Positive output of relay RLY1 is connected to the red positive of the binding post terminal.

Testing the system

- To test the system, charge the receiver capacitors and the transmitter capacitor. During charging, all switches on the receiver and transmitter should be in the off position.

- Turn the transmitter safety key switch to the "on" position. The red LED will light, and indicate that the firing button on the transmitter is ready to push.

Testing the system, continued

◆ Turn on the receiver with the safety-on switch.

◆ Depress the firing button on the transmitter.

If you hear the relay contacts closing in the receiver, the circuit is operating properly. There may be some relay chattering at first but it will stop for a solid contact. If it does not, you may have to charge the receiver supercapacitor C1 more.

Using the system

This system at minimum has about a 300 foot range.

You will need a two conductor wire four to six feet long with alligator leads on one end, and ring or spade terminals on the other.

Make sure that you have coded or depressed the DIP switches in both receiver and transmitter for the same code.

◆ Check all receiver and transmitter switches to be sure they are off.

◆ Connect igniter to rocket motor.

◆ Connect the alligator leads to the rocket igniter.

◆ Connect the ring, spade or plug terminals to the black negative and red positive terminals on the receiver output. When doing this make, be sure you

are out of harm's way in respect to the rocket and rocket motor exhaust.

◆ Turn on the safety switch on the receiver and proceed to move away from the receiver rocket launch pad with the transmitter in hand.

◆ When you have reached your firing position, turn on the safety key switch on the transmitter.

◆ When you are ready to fire the rocket, depress the firing momentary push button switch.

◆ When you see ignition occurring, release your finger from the firing button and turn off the key switch immediately to conserve power.

◆ After launch, walk back to the launch pad area and turn off the receiver safety switch to conserve power.

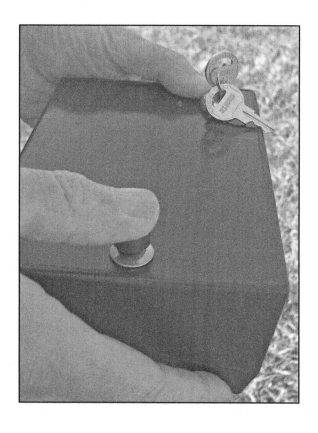

If ignition fails, always turn off transmitter, wait a few minutes and then turn off the receiver safety switch and disconnect the wire from the binding post terminals. Then, check connections with alligator clips on igniter.

Never adjust igniter when it is connected to the receiver box.

Always remove the transmitter key after turning the transmitter off and keep the key on your person.

Always turn receiver off before checking for malfunctions.

This system is reliable, simple, and works quite well. We have used it for many launches. It will work forever without ever needing a battery and there is no need to string long lengths of snarly wire to and from the igniter and controller.

The system works well with Estes igniters, but other igniters can be used or made to suit your needs. They would, of course, have to be tested with the system for compatibility and tweaked.

For making igniters from Nichrome resistance wire, we have found that #30 wire works well. You can, of course, try other gauges to suit your needs.

The capacitors can also be charged from a crank generator in the field, if the weather should turn nasty on you. Before switching to the supercap system, we used crank fire, as its called, to fire most of our rockets as a greener alternative than using batteries. We still take the generator out to the field, just in case, to have a reliable backup to charge the supercaps.

As with any wireless device used for ignition, caution is always needed. Although the system is coded to help to minimize other ambient signals from causing ignition, always proceed with caution and safety first in mind. As with any experimental device, you are responsible for its use and you must conform to all relevant safety laws required by your AHJ (Authority Having Jurisdiction). This is an experimental device and as such is not considered safe by any standards. We do not encourage the building or use of this device by anyone not willing to fully assume the risks and liability.

Supercaps and the Home PV System

Most photovoltaic systems that have battery banks can be improved with the addition of supercapacitors. Appliances that have motors can draw excessive current from the batteries every time the motors start up. Supercapacitors can extend the total life cycle of the batteries by preventing these demands on the system from drawing down the batteries as quickly.

Photovoltaic systems typically use flooded or sealed lead acid deep cycle batteries for energy storage. Some grid tied systems do not employ battery backup, but some do in case of grid failure. Flooded lead acid batteries come in a variety of configurations, the most common being 2 volt cells, 6 volt batteries and 12 volt batteries. Sealed lead acid are usually 6 volts or 12 volts.

Either type can be used in a supercapacitor and battery hybrid system. Other less used options for a supercapacitor and battery hybrid system are nickel cadmium, nickle metal hydride, lithium ion and nickel iron batteries. Each have their own charging peculiarities and are not preferred options due to cost and other factors.

The most commonly used deep cycle batteries are flooded lead acid such as Trojan T105s. They're popular because they're readily available, relatively inexpensive, and are good quality. Trojan T105s are 6 volt, 225 amp-hour batteries and in most systems are connected in series to form a 12 volt battery. These 12 volt pairs are then connected in parallel

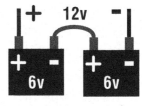

Series connected batteries

Two pairs of series connected batteries, connected in parallel

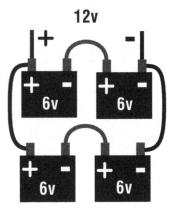

with other 12 volt pairs for more amp-hours. Although this type of series parallel arrangement is the most common there are 24 volt and 48 volt series parallel configurations that are used. Supercapacitors can be connected in parallel to any of these battery arrangements.

Our example uses four 6 volt flooded lead acid batteries connected in series parallel to provide 12 volts for our system operating voltage. This DC voltage can be used directly to power 12 volt DC devices, or it can be routed through an inverter to provide 120 volts AC.

Designing the supercap bank

12 volt photovoltaic systems with flooded lead acid batteries are generally charged at around 14.5 volts and equalized at 15.5 volts. Thus the supercapacitor bank for such a system must be able to handle at least 15.5 volts. Here, we used six 2.7 volt Maxwell 650 farad supercapacitors connected in series. They can be safely charged to 16.2 volts. Unlike batteries, supercapacitors can be charged to less than their specifications and not degrade.

A supercapacitor bank must be rated to handle at least the highest voltage that your system will be subject to. The voltage from the panels going to the batteries and supercapacitors is regulated by the charge controller in your system. If you are using sealed lead acid batteries, it is important to have a very good charge controller that does not deviate above the stated charging voltage because charging at higher than the rated voltage will quickly reduce the life cycle of the batteries.

Most systems will use 6 or 12 volt batteries. It is possible to use large 2 volt cells. The 2 volt cells are the best for this application as you can apply the supercapacitor in parallel to the terminals of each cell and then connect the cells in series for your final voltage requirements. This arrangement does not need active or passive balancing and thus reduces energy loss through controlled leakage current. Although available, 2 volt cells are much more expensive than 6 or 12 volt batteries. For this reason they are not employed in most photovoltaic systems. You can build 2 volt cells, and if you have an interest, complete information can be found in the *Battery Builders Guide*.

Integrating the supercaps and the battery bank

For our project, the positive and negative output of our 12 volt battery bank is connected to the positive and negative terminals (in parallel) of the supercapacitor bank. This particular supercapacitor bank employs active balancing. A Maxwell integration kit was used which includes active balancing circuitry, bus bars, nuts, spacers, washers, battery posts (optional), wiring and plastic placeholders. This kit is handy and relatively inexpensive. Another option would be to use passive balancing with resistors and make your own bus bar connectors, etc.

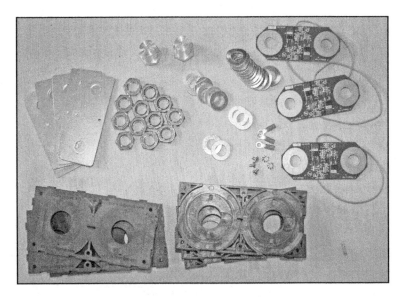

Maxwell integration kit

It is important for any hybrid supercapacitor and lead acid battery system to have an enclosed supercapacitor bank to avoid the corrosive effects of electrolyte spray.

Battery banks emit electrolyte into the surrounding air and this can be very corrosive to the aluminum supercapacitor bodies, terminals, and electronics. You will need an enclosure to protect the supercapacitor bank. Although not professional looking and lacking in aesthetics I have used plastic cling wrap to cover and enclose supercapacitor banks as it has good resistance to electrolyte and is a quick and inexpensive answer to the corrosion problem. However, for most installations you will want to provide a sturdy enclosure.

For this example we used a NEMA rated waterproof enclosure. It is made of plastic and does not have ventilation for heat dissipation. Heat dissipation was not a concern for this application

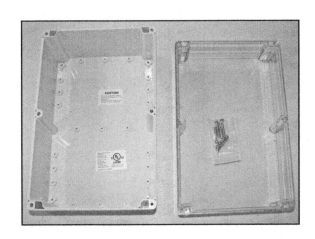

because our current needs were low. However, extremely high, and prolonged current draws will create heat within the supercapacitors. If you anticipate such heavy use ventilation should be added by drilling holes in the box to create air flow for excessive heat to dissipate.

Constructing the supercapacitor bank

For this particular example we used six Maxwell 650s connected in series with active balancing accomplished through the Maxwell integration kit.

The heart of the integration kit is the active balancing circuit board, see photo at right. The three boards that come with the six supercapacitor integration kit have LEDs on them that light up when balancing is occurring between the supercapacitors.

The integration kit includes a manual with photos detailing the assembly steps for constructing the supercapacitor bank.

Active balancing circuit board

Basic assembly consists of placing the supercapacitors in their holders and attaching the circuit boards and bus bars and attaching bus bars to circuit boards with a

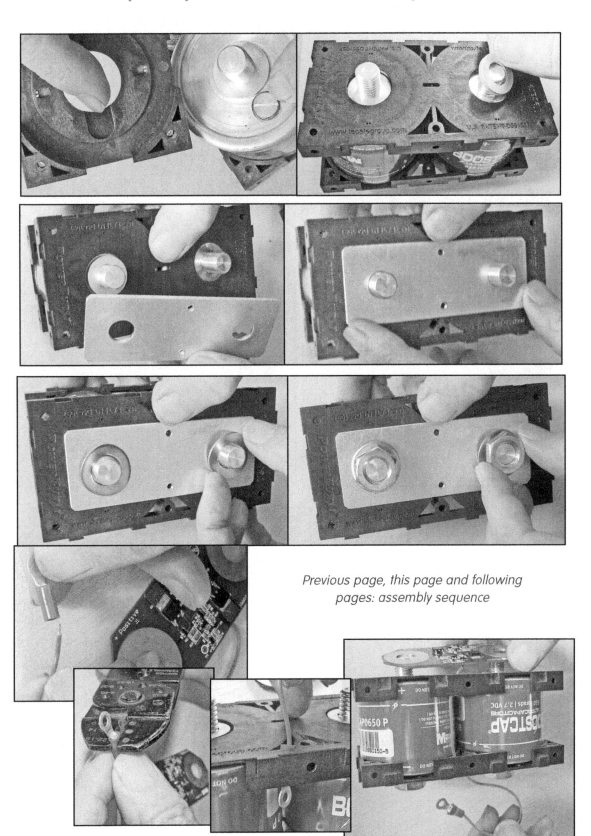

Previous page, this page and following pages: assembly sequence

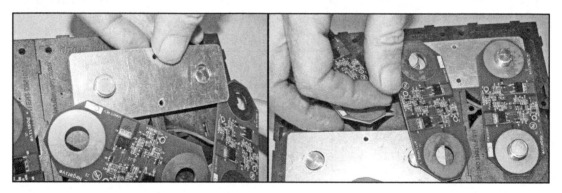

wire. The plastic assembly is connected together via slots to form the finished super-capacitor bank. You have the option of ordering the integration kit with battery type terminals included. These are not necessary as you can use the nuts provided with ring terminal connectors, but they are a good option if you wish to use standard battery cable connectors as well. The accompanying photos show the construction sequence.

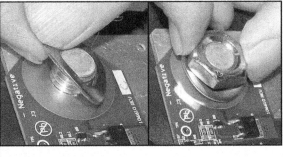

Constructing the supercapacitor bank enclosure

The enclosure for the supercapacitor bank needs ports for the input and output wires connecting the supercapacitor terminals to the battery terminals. The input and output ports are made tight by installing a gland or rubber slot. This could also be done by installing a connector in the wall of the box to provide a connection between the supercapacitor bank on the inside with the wires from the battery connected to the outside.

When making large size holes in plastic boxes, first drill a small pilot hole and then ream out the hole to final size hole with a hand reamer. For this unit, one port is for the positive and one port for the negative wires that lead from the supercapacitor terminals to the battery terminals.

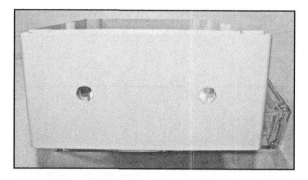

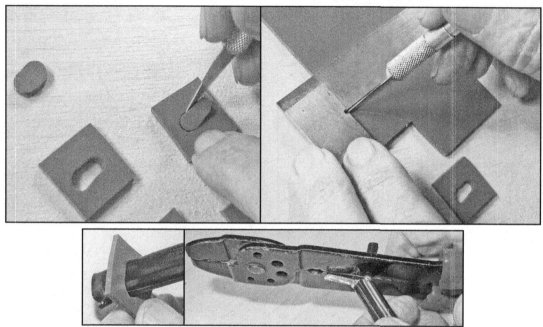

To make the rubber slots for two entry ports, measure and cut four silicone rubber pads. A hole is then cut in each of the pads to make a space for the wire or cable to go through. Test fit the wire to see that it is a tight but workable fit.

You can purchase battery/inverter/welding cable cut to size and fitted with ring terminals, or you can cut your own cable and crimp and solder the terminals yourself. Here, we used both conductors of 10 AWG zip cord for each lead wire.

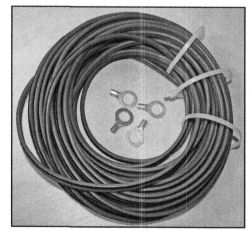

The zip cord was cut to the length needed and covered with electrical tape for its full

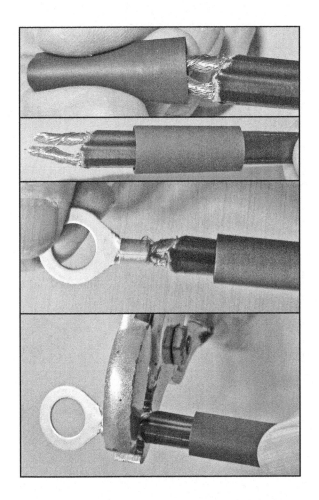

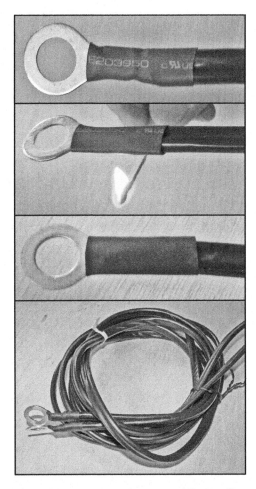

length to further protect it. The electrical tape was not necessary but it adds to the longevity of the cable. The next task was to strip the zip cord, crimp a ring terminal on one end of each cable and cover each terminal crimp with heat shrink tube.

The four silicone pads are epoxied on both sides of the holes. This prevents electrolyte spray from entering the enclosure through the ports.

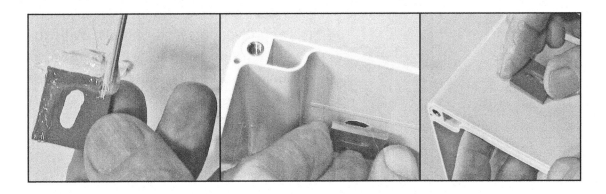

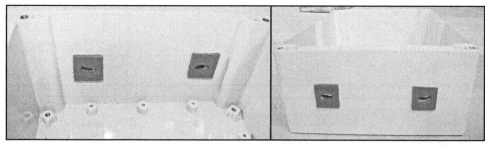

The box we used had molded stand-offs on the inside which hindered proper seating of the supercapacitor bank. To provide a stable platform, a piece of Plexiglas was cut to fit on top of the standoffs inside the box.

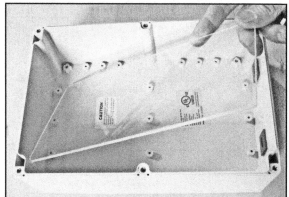

The next step was to thread the uncrimped ends of the wires through the holes in the box from the inside out. After the wires were threaded through the box, ring terminals were crimped to the ends that would connect to the battery bank.

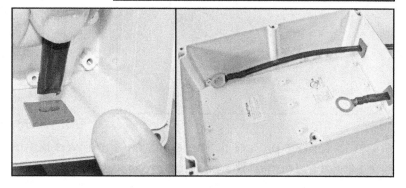

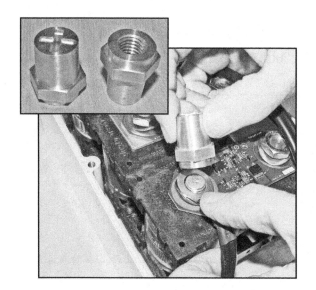

The supercapacitor bank is slipped into the enclosure and the ring terminals connected to the positive and negative output terminals. With the integration kit you get the choice of using battery type terminals or hex nuts. We chose to use the battery type terminals for a variety of reasons but you can use either to secure your ring terminals.

The next step was to place the six screws to hold the bottom and top portions of the box securely together.

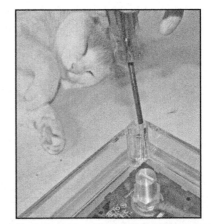

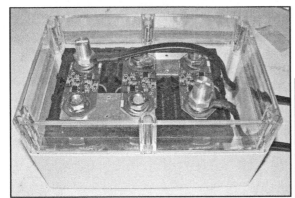

The supercapacitor bank is now ready to be connected to the battery bank.

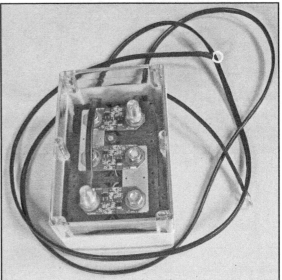

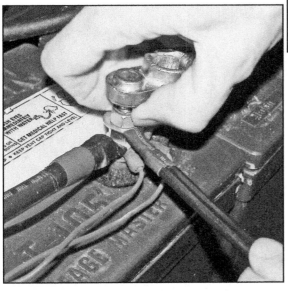

The connection between the supercapacitor bank and the battery bank is parallel, so the positive terminal of the supercapacitor bank connects to the battery bank positive output, and the negative terminal of the supercapacitor bank connects to the negative output terminal of the battery bank.

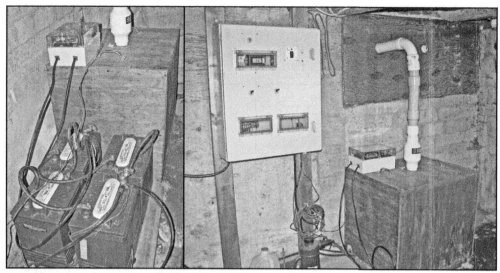

For us, it was convenient to locate the supercapacitor bank on top of the battery box.

Parts and suppliers

Ultracapacitors, 650 Farad 2.7 volt	Six Maxwell BCAP0650-P270 with threaded terminals. *Tecategroup*
Integration kit	Maxwell Power Burst Kit for six cell modules *Tecategroup*
Box	Bud Industries NEMA 4X sealed polycarbonate, light gray body, clear cover, 9.45"x 6.30"x 4.72". Supplier *Allied Electronics #736-2941*
Zip cord	10 AWG 30 amp. *The RF Connection*
Battery and inverter cable	*New England Solar* or *Backwoods Solar*
Shrink tubing	Electronics, electrical, hardware, or automotive stores
Ring terminals	Crimp on. Hardware, electrical, automotive, or electronic stores
Silicone rubber sheet	Hardware store, plumbing section or *McMaster Carr*
Glands (rubber protective covers)	*Allied Electronics*
Weather tight connectors	*McMaster Carr, New England Solar*
Wire, Cable, Terminals, Connectors, Shrink Tube	*Wiring Products*
Aluminum stock for bus bars	Aluminum alloy 6061, dimensions as needed. *McMaster Carr*

Parts numbers and suppliers have been provided for your convenience; however, suppliers may go out of business and parts numbers may change. All parts listed are available from multiple suppliers.

Solar Supercapacitor Powered LED Lighting

Light emitting diodes (LEDs) are excellent sources of illumination and are beginning to replace both incandescent and fluorescent bulbs for general lighting purposes. Their energy efficiency makes them the preferred choice to use with renewable energy sources, and thus solar supercapacitor operation.

Solar supercapacitor LED lighting systems can be configured to be portable, as in flashlights and lanterns; or stationary, as in household or workplace lighting fixtures. A wide variety of LED light bulbs are available with a standard screw-in base so that they can easily replace conventional bulbs. LEDs run on DC current and the bulbs have a small inverter in their base that provides DC to the LEDs.

As an example of LED solar supercapacitor lighting options, consider repurposing a simple desk lamp to LED supercapacitor operation. The simplest option is to purchase an LED light bulb, screw it into the lamp socket and provide the power with something similar to the universal power supply through the AC inverter attachment (see page 46). This way the desk lamp can run from supercaps, or from the household outlet when necessary by simply unplugging the lamp from the supercap power supply and plugging it into a household AC outlet. This requires no modification or alteration to the lamp.

Running the lamp on DC

Another option would be to fabricate a screw based LED that would fit the lamp socket and run on DC without incurring the energy losses of the conversion from the inverter. There are several ways to accomplish this. It could be powered by a power supply that has a Micro Puck boost converter in it. The universal power supply would be modified by adding a switch to route the 2.5 volt supercaps through the boost converter, then to a regular household outlet on the outside of the power supply box.

Another option is to install the Micro Puck boost converter into the screw assembly. This allows operation of the LED directly from a 2.5 volt supercap power supply without adding a switch to the universal power supply box.

You could also insert both supercapacitor supply and boost converter in the base of the lamp with an input jack for a solar panel.

There are many variations on a theme which can be applied. The two examples detailed here will get you pointed in the right direction.

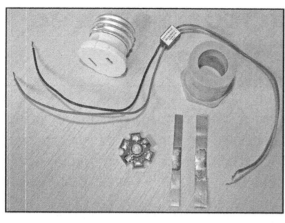

LED light adaptors

Materials needed to make an LED light adaptor

Washer (for simple adaptor only)	Rubber, plastic, or fiber, $1/8$" thick, and about 1" outer diameter, $3/4$" inner diameter. Hardware store
Copper strips, two	Cut from copper flashing or thick copper foil, one 2" by $3/16$"; one 2"x $1/4$" wide. Hardware store.
White LED	One Luxeon Star LED LXHL-MW1D. *LED Supply*
20 gauge wire, stranded	Two pieces, $1\,3/4$" long or less, $1/4$" stripped on each end
Lens optics (optional)	*LED Supply*
Goose neck lamp	
For adaptor with internal Micro Puck:	
Micro Puck boost converter	Micro Puck 2009A, 350 milliamp boost converter. *LED Supply*
PVC reducer	One grey $3/4$"x $1/2$" reducer. Electrical section of hardware store

Parts numbers and suppliers have been provided for your convenience; however, suppliers may go out of business and parts numbers may change. All parts listed are available from multiple suppliers.

Tools

- Glue
- Soldering gun
- Soldering iron
- Wire cutters
- Stripper
- Solder

Constructing a simple adapter

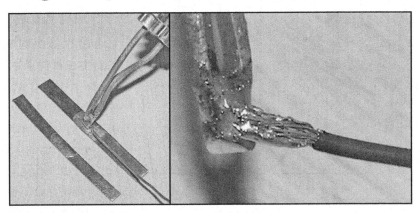

Cut two copper tabs to serve as plug inserts. Tin one side of each insert in the middle with a soldering gun. Tinning is simply adding a very small amount solder to the surface of the inserts.

Next, cut the wires, strip both ends of each wire and tin the ends. This can be accomplished by melting some solder on the tip of the gun and wiping it across the stranded wire until the strands are coated.

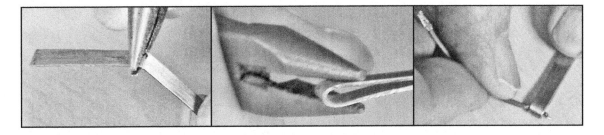

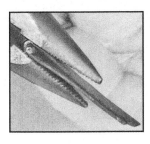

With needle nose pliers, bend the inserts at the center, tinned side inward.

Insert one end of a stripped wire into the bend of each of the inserts.

With pliers, compress the bend with the wire sandwiched between the copper layers.

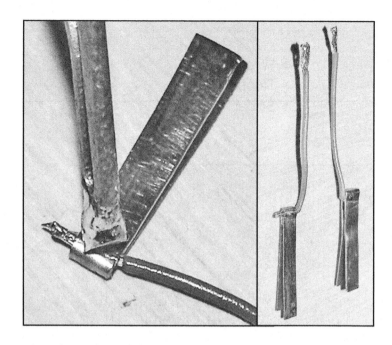

Apply the soldering gun to each bend. This will melt the solder on the insert and the wire and bond them to provide a good electrical connection.

Next, solder the wires from the inserts to the Luxeon star. Do not use a soldering gun for this operation. Instead, use a lower wattage soldering iron. The soldering gun is fine for connections such as the wire and inserts where more heat is needed due to the thicker metal, but it is not good for electrical components which can be destroyed by excessive heat. All you need to do is solder one wire to the positive pad and one wire to the negative pad on the star.

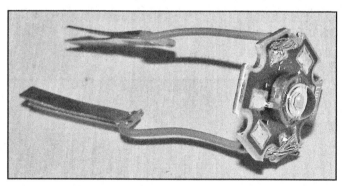

Bend the wires downward through the indentations on the scalloped edge.

Glue a non-conductive washer to the receptacle face with epoxy. Let this dry for twenty four hours.

Next, insert the copper tabs into the receptacle holes making sure they are fully seated and do not extend beyond the receptacle surface.

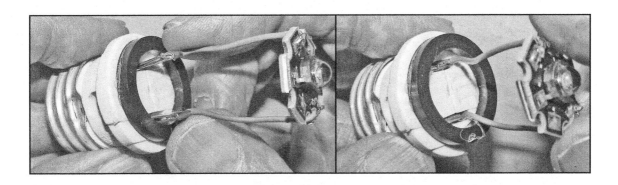

Push downward, keeping the wires inside the ring of the washer. Push fit until the bottom surface of the star sits on the surface of the washer.

If you let go of the Luxeon, notice that it springs away from the washer due to the pressure from the bent wire. Coat the top of the washer with epoxy and seat the star again on the washer. Apply a rubber band to hold the star to the surface of the washer while the glue is bonding.

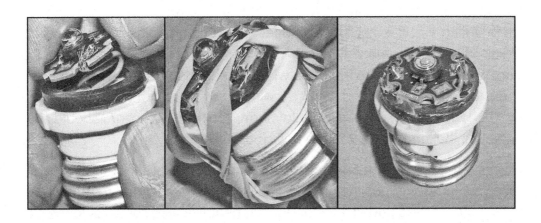

Let it dry for 24 hours, then remove the rubber band.

You can also add lens optics for different illumination patterns

To use, screw the adaptor into the lamp. Connect to a supercapacitor supply with a Micro Puck output to power the lamp.

Warning

The LED module that you have just constructed works on DC only. Never plug it into your home AC outlet. This modified lamp is solely intended to be used with a 2.5 volt supercapacitor supply. Either attach a tag to the plug or place a warning label on the lamp stating that it is not to be plugged into a common household socket. Tags and labels, however have a way of coming off or may not be understood properly by someone who decides to use the lamp when you are not present. For this reason it is advisable to cut the plug off and replace it with a connector that will only fit your power supply. This way you can avoid accidental harm, shock or fire hazard should someone decide to use the lamp while you are not around.

Also note that the Luxeon Star is polarity sensitive. Be sure that the connector you attach to your lamp cord has the positive and negative wires marked. The positive wire connects to the positive pad of the star and the negative wire connects to the negative pad of the star. If they are connected incorrectly you will burn out and destroy the LED when you plug into the power supply.

If you are not using one of the power supplies in this book for either of these modified lamps, be sure to include a 1 amp fuse in the lamp's circuit.

Adaptor with internal Micro Puck

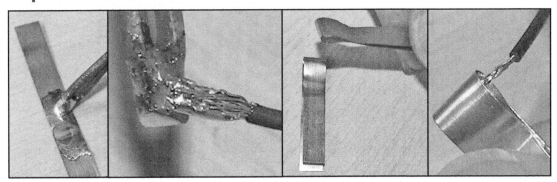

Cut and tin the copper inserts as done for the other adaptor.

Strip and tin the four leads of the Micro Puck.

Bend the copper inserts as for the other adaptor

Slip each of the black and red input leads of the Micro Puck into a copper insert.

Press firmly with pliers and apply the soldering gun to bond the wires to the inserts.

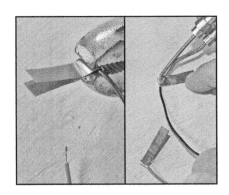

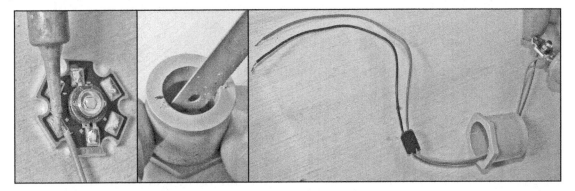

Solder the orange positive output wire to a positive pad on the star. Solder the green negative output wire to a negative pad on the star.

To prepare the reducer for inserting the star, cut two grooves opposite each other on the inside edge on the end of the reducer that has the round rim. These grooves are for the wires from the Micro Puck.

Thread the Micro Puck wires through the reducer and test seat the star to see if the wires fit well in the grooves and allow the star to rest firmly on the surface edge of the reducer.

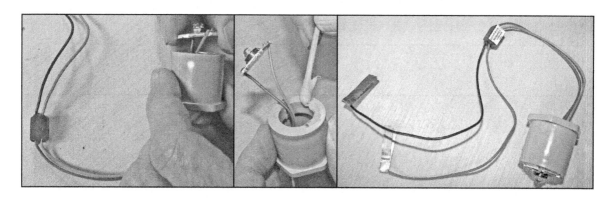

If the star seats well, apply epoxy to the round rim of the adaptor and seat the star on the reducer to bond. Let this dry for twenty four hours.

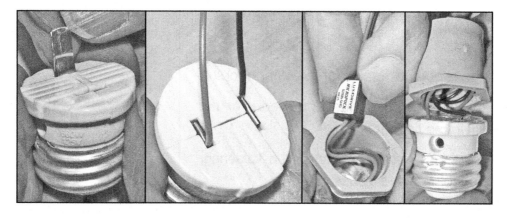

Insert the copper tabs into the plug adaptor.

Stuff the wires and the Micro Puck into the reducer.

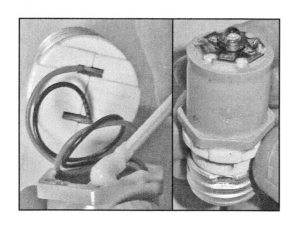

Coat the hexagonal end with epoxy and bond the adaptor to the reducer. Use a rubber band to hold components in place. Let this dry for twenty four hours.

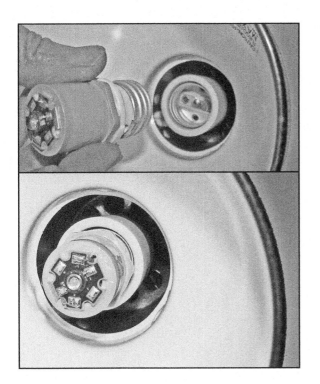

To use, screw the adaptor into the lamp and connect directly to a 2.5 volt supercapacitor bank.

Warning

Never plug the modified lamp into a house hold AC outlet, be sure the lamp's positive and negative wires connect correctly to the Luxeon Star, and include a fuse in the circuit if the lamp is not used with the supercap power supplies in this book. Review warning details.

Solar Supercapacitor Powered Lasers

Lasers are useful devices with many applications. Although there are many types, the most commonly used is the diode laser. Laser diodes can be found in CD players, laser printers, laser cutters, laser pointers, supermarket scanners, carpenter levels, as surgical instruments, and in a myriad of scientific devices.

I use diode lasers quite frequently in the field and in the laboratory. Most diode lasers operate on 3 to 6 volts DC, and I find using a universal power supply with a modified Recoton attachment works quite well for my purposes. I use the 15 volt series connected bank and the Recoton (see page 45) to deliver the voltage needed for a particular laser.

One of the best sources for lasers to modify are laser pointers as they are readily available and relatively inexpensive.

Laser modules can also be purchased. These are simply the diode, driver and optics contained in a metal case. The metal case not only holds the parts but acts as a heat sink.

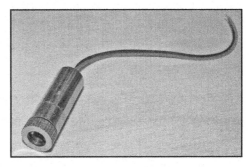

Laser module with adjustable focus

Modules are easy to work with. All you have to do is connect the input wires to your power supply output. Most modules come with focusing optics which generally consist of a plastic lens set in the case end. The lens screws in or out to focus the beam.

Another option is to purchase a diode, driver, lens, and module case as separate items and design a module.

Modifying a laser pointer

With a few simple modifications, laser pointers can be powered with a solar supercapacitor power supply. In theory, all you need to do is to remove the

Laser driver, and diode

battery supply and connect the input leads to the solar supercapacitor supply. In practice you have to substitute a battery shaped space filler that will take the place of the batteries and connect the power supply to the driver.

This is easily accomplished by making an insert with leads. The diameter of the insert must fit the particular pointer you are modifying.

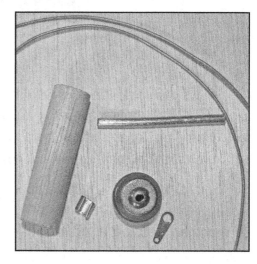

Wooden dowel rods can be used for inserts. Unscrew the end of the pointer and see what size dowel rod fits snugly inside the pointer. This will generally be the diameter size of the batteries used in the pointer. Most pointers use AA, AAA, or CR2 type batteries.

Note the number of batteries in the pointer and cut a piece of the dowel rod the exact length that the batteries take up in the pointer holder. For instance, one laser unit I have uses two CR2 batteries. I lay them end to end, place

Parts needed to make the pointer insert

a dowel rod next to them, mark the length and then make the cut. This rod will take the place of the batteries and must be the same diameter and length.

Once the dowel is cut to length, mark the center of the dowel and drill a hole in the center of the rod all the way through to the other end. This hole must snugly fit a 6 or 8 gauge solid wire. You can use a drill press for this, or a hand drill in a jig for an accurately placed hole.

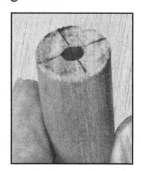

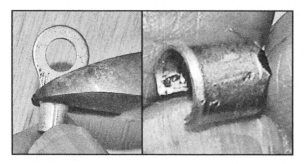

Next, cut off the crimp part of a crimp on ring connector.

Measure the length of the cut crimp part and drill a hole that deep on one end of the dowel rod, centered on the prior drilled hole. The diameter of the hole must be the same diameter or a tad larger than the diameter of the crimp piece.

Cut a piece of 20 gauge wire about 6" long, strip the ends and tin them.

Cut a length of 6 or 8 gauge solid copper wire about ⅛" longer than the length of the dowel piece.

Open the crimp piece a bit with needle nose pliers.

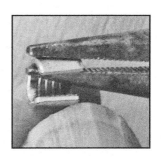

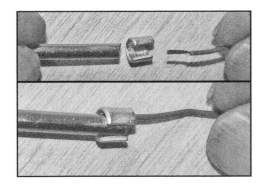

Insert the copper 6 or 8 gauge wire and one of the stripped ends of the 20 gauge wire into the crimp piece.

When seated, apply pressure with pliers or crimping tool to mechanically connect the wires firmly together.

Apply a soldering iron to the crimp piece to melt the solder on the 20 gauge wire.

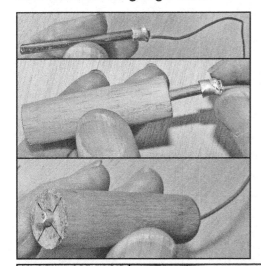

Push this assembly into the dowel rod and seat it.

For most laser pointers this will be your positive lead but in some pointers this will be the negative lead depending on how the batteries are required to be inserted via the directions that come with the pointer. Pay careful attention to these polarities. Use a red wire for the positive and a black wire for the negative, or mark the wires to avoid confusion. If you hook up the driver and diode the wrong way you can destroy the diode.

Next, solder a 6" piece of 20 gauge wire to a solder lug.

Place the soldered lug on the end of the dowel and bend the tab over the edge of the dowel. This will be the negative lead unless, as noted above, the batteries were inserted backwards, in which case this would be the positive lead.

The bent tab will make contact with the inside of the case. In most pointers the case is electrically conductive and acts as a path in the circuit. Although this is the norm, be sure to inspect and understand what and where the conductive paths are for your particular pointer, and attach the negative and positive leads appropriately.

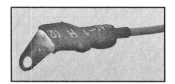

After the tab is bent, thread a piece of shrink tube over the wire and onto the terminal. Heat it to form an insulated surface that will keep this wire from shorting out against the other wire in the assembly.

Open the pointer case, and slip the dowel rod in. Feel to ensure that the tip of the wire is contacting the spring firmly on the driver board. In most laser pointers a spring is the contact for the battery terminal, and if the wire in the center of the dowel is too small in diameter, it may go through the center of the spring without making contact, or the contact will be erratic at best.

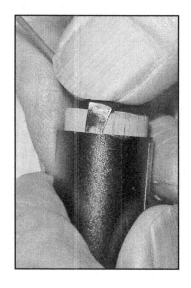

If it is not contacting, mushroom out the wire by pounding it a bit with a hammer to the point so that the wire will push against the spring rather than go through it. If you use 6 gauge wire instead of 8 gauge wire you will probably not have to bother with such modifications. It all depends on the diameter of the hole at the end of the spring. In most laser pointers you can unscrew the driver and diode assembly and remove it for inspection to see what the internal contacts are and build the insert accordingly.

The solid conductor wire at the center of the dowel was cut so that it protrudes about 1/8" from the surface of the dowel. This creates wiggle room. If you do not need

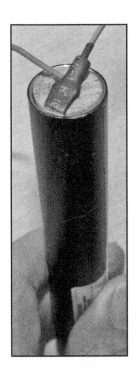

the extra wire length, you can file or grind it down so that it abuts (is flush with) the surface. If it fits well with the protruding piece, leave it.

Next, insert the other conductor between the dowel rod and the case. This should be a firm fit. The wood is soft so it will slightly compress into the space for a snug fit.

Finally, an end cap is needed for the pointer. Here you will have to use your own creativity. More than likely you will

not be able to use the original end cap for the pointer and will have to find something that fits and encloses. You can also leave it as is without an end cap.

For the laser in the example, I used an end cap from a Recoton plug and fuse holder. It was exactly the right diameter. I inserted a rubber grommet to reduce the hole for the thinner wires.

There are hundreds of parts at your local hardware store that will be perfect for this. All you have to do is go to your local hardware store and spend hours musing on a variety of parts. This can be fun or not depending on your perspective.

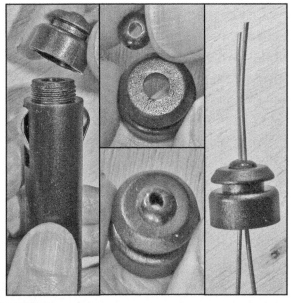

For this particular laser I connected the end cap with a piece of electrical tape to keep the option of using the laser pointer without the modification. The modification can be easily dismantled to restore the laser to its original configuration.

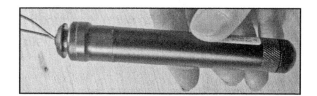

Laser platform

Most scientific experiments involving lasers need a stable steady laser beam that can be directed and stationed in a variety of positions. To accomplish this, I built a platform that can be mounted on a tripod. The platform has a V block with a screw clamp that can hold any diameter pointer in place. It also dampens vibration.

A 1/4"-20 perforated base nut attaches the platform to the tripod.

The platform has a jack for power input and two Fahnstock clips that hold a resistor. The resistor is a safety precaution to avoid burning out expensive diodes in case I thought that a diode driver might not provide the proper current limitation. For the most part you will not need this feature since most drivers in laser pointers and modules will limit the current properly to the diode.

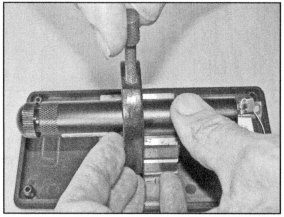

I also use the resistor to purposely limit the current below specifications so that if I can run the laser with a weaker beam if I want to. Another way to do this would be to hook in a potentiometer to be able to have variable resistance. If you have a fixed resistor, the Fahnstock clips can be attached with Velcro or epoxied into place. Any resistor must be watt rated for the current draw and voltage required for the laser.

The platform made was a project box top I had on hand that was about the right size. A platform can be made from many types of materials according to your needs.

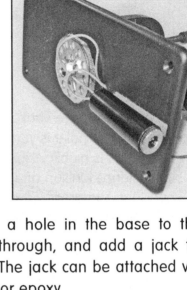

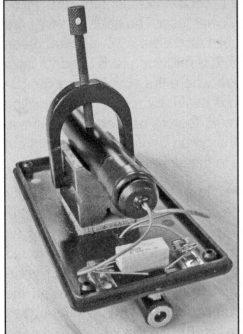

Drill a hole in the base to thread the wires through, and add a jack for power input. The jack can be attached with either Velcro or epoxy.

The V block can be secured to the platform with either Velcro strips for temporary attachment or epoxied for permanent attachment.

Solar panel

The solar panel for charging the supercapacitor bank is a 2.5 volt system panel and is mounted on a tripod so that I can track the sun for maximum output to charge the supercapacitor bank.

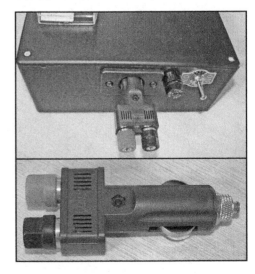

I used the universal power supply with a binding post plug attachment which can directly power the laser through the panels during the day or from the charged supercapacitor bank at night.

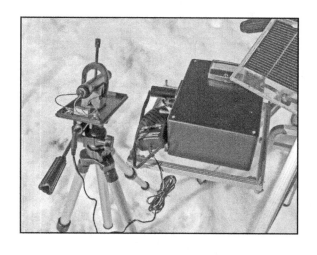

There are numerous experiments one can perform with this setup. Our experiments have included laser voice and data communications, long distance ignition, and signaling and rescue applications.

Laser diodes can provide a variety of wavelengths and thus different colored output. Some beams are invisible, some are barely perceptible and some are very easy to see as the eye is more sensitive to certain wavelengths and not so sensitive to others. For instance I have a red 200 milliwatt laser. The beam is very hard to see although the focus point is very bright and hot. I also have a 50 milliwatt green laser with a beam that is highly visible at night. The 200 milliwatt red is used for such experiments as distance ignition and communication where the visual beam component is not important. The green 50 milliwatt beam is useful for rescue and signaling operations as the visual beam component is the main feature needed. Astronomers like the green beam lasers to point out stars in the night sky.

One laser experiment was a solar supercapacitor setup for distance ignition of black powder. Black powder has an ignition temperature of around 630°F, depending on the exact composition of the powder. We tested the ignition capability of a 200 milliwatt red diode laser set in the V block and powered by a solar supercapacitor supply with an attached Recoton. The ignition photos are time lapse taken from video. The distance was about 16 feet from the laser head.

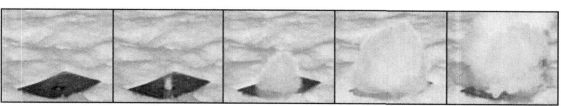

The beam in the photo at left is a 50 milliwatt green laser diode at night for signaling and rescue experiments, powered by a solar supercapacitor supply.

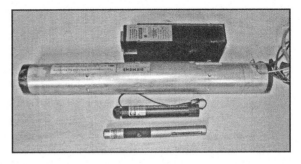

Top to bottom: two helium neon lasers; two diode lasers.

Another type of laser is the helium neon gas laser. These require a high voltage power supply which is usually contained within the laser module or comes with the tube when purchased from surplus suppliers. They are not as common as they used to be as they have been replaced for most functions by diode lasers.

Any type of laser can be powered with a solar supercapacitor system. You simply need to look at the power requirements for the device and build your system accordingly. Some laser power supplies require 12 volt DC input, and others, 120 volt AC input. In either case, the universal power supply with or without AC inverter attachment can meet those needs.

Safety warning

When experimenting with lasers, you must have eye protection. Laser protective glasses are available from a wide variety of sources (see the list, next page) and are designed specifically for the wavelengths that you will be working with. Red lasers will require one set of glasses while green lasers or infrared lasers will require a different set of glasses. Common lasers such as found in key chain lasers and pointers that are 5 milliwatts and under do not pose as much of a hazard unless beamed directly into the eye. But anything above that should be handled with extreme caution as eyesight is precious and it takes less than a second to destroy it with a laser. Both infrared and ultraviolet lasers can pose significant threats to eyesight. To understand what the best safety options are for the power and wavelength you will use, refer to American National Standards Institute ANSI Z136 for the US standards, and International Electrotechnical Commission IEC 60825 for international standards.

Parts and suppliers

For more information about lasers, check *Laser Pointer Forums.* This forum contains the latest updated information, including where the best prices are to be had for a variety of laser types and parts. Another good site is *Sam's Laser FAQ.*

Lasers, laser pointers	Power output depending on needs There are many suppliers of laser pointers and other types of lasers such as *Deal Extreme, Amazon, Meredith Instruments* You can also find lasers and laser parts on popular internet auction sites.
Laser safety glasses, lasers	*Dragon Lasers*
Camera tripods	Available at a variety of outlets
Supercapacitor power supply	See *Universal Power Supply*, p.45
Solar panel	See *PV panel voltage needs*, p.11
Vehicle power adapter plug	With banana jack binding post. *Radio Shack* 270-1521
Modification materials for laser pointers	
Velcro strips or tabs	Sewing shop, hardware store, other outlets
Binding nut	Perforated base nut 316 stainless steel, $1/4$"-20 thread size, *McMaster Carr* 98007A029
Base/platform	Can be either wood or plastic. I used the top of a plastic project box
Fahnstock clips	Two, available at most electronics parts suppliers. *Ocean State Electronics* 2090
Dowel rod	Diameter will depend on inner diameter of the laser pointer. I used a $5/8$" for this example. Hardware store
Solder lug terminal	(Copper metal foil can be used instead) *Allied Electronics*, other electronics stores)

Shrink tube	Electronics store
20 gauge wire	Electronics store
6 or 8 gauge wire	Used for ground wire. Local hardware store, electrical section.
V blocks with clamps	*Micro Mark* 14255
Resistors	Fixed resistors or variable resistors (potentiometer/rheostat) used with laser diodes must be rated for the proper wattage. For example, for a laser pointer that consumes 300 millliamps at 6 volts that would be 1.8 watts, so the resistors must be rated at 1.8 or above. Generally I use 5 to 10 watt resistors. Any electronics store
Various connectors	Electronics store
Coaxial DC power jack	Size M, *Radio Shack* 274-1577
Modified Recoton	See *Modified DC to DC converter, p.41*

Parts numbers and suppliers have been provided for your convenience; however, suppliers may go out of business and parts numbers may change. All parts listed are available from multiple suppliers.

Other Solar Supercapacitor Applications

High voltage

As an experimenter I have many uses for high voltage devices including x-ray tubes, particle accelerators, metal deposition, Tesla coils, RF communications, micro-waves, ionizers, plasma experiments, lasers, and electron microscopes.

There is a variety of power supplies one can build or purchase on the surplus market to meet your specific high voltage needs, including neon transformers, solid state supplies, induction coils, Van de Graaff generators and voltage multipliers (Marx generators).

High voltage power supplies run on DC or AC current. Either can be powered with a solar supercapacitor power supply.

Van de Graaff generators have either a DC or AC motor to move the belt that places the charge on the sphere. Any 12 volt system supercapacitor bank can do the job directly for DC, or with the inclusion of a small inverter, run an AC motor.

Other types of supplies can run off either AC or DC so it is just a matter of supplying the right voltage and type (AC or DC).

I have found that most, if not all, of my high

Several high voltage supplies with supercapacitor bank, inverter, and connectors

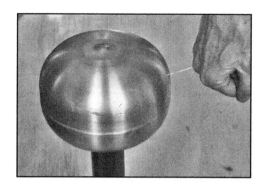

voltage needs can be met with solar supercapacitor supplies since most of the equipment is used for very short run times.

Van de Graaff generator

Surplus high voltage supply with inverter, supercapacitor bank and connectors

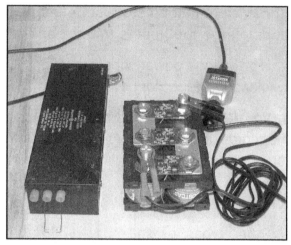

A good source for information about building and using high voltage power supplies is Gordon McComb's book *Gadgeteers Goldmine*. There are plenty of plans and information about high voltage supplies to be found on the internet. High voltage supplies are available from *Fair Radio Sales, Surplus Sales Of Nebraska, Information Unlimited, Science First, Ramsey Electronics* and many others.

Solar hydrogen fuel cell systems

Supercapacitors can be used in parallel connection with electrolyzers and/or fuel cells to aid in a smoother delivery of either gas output or voltage generated from the fuel cells.

Solar powered electrolyzers reduce output with quickly changing cloud transits. A supercapacitor allows smoother, more consistent gas production.

Coupled in parallel with fuel cells, supercapacitors provide a buffer for a fuel cell or fuel cell bank when current surges are needed to start electromagnetic devices such as motors.

Supercapacitor connected in parallel with fuel cell

Thermoelectric applications

Thermoelectric Peltier type modules can be powered by supercapacitors to produce either cold or heat. Supercapacitors can also be charged by a bank of Peltier modules with the application of heat to the modules. Although not a particularly efficient application, it is interesting to experiment with. Peltier modules can be purchased from *American Science & Surplus*.

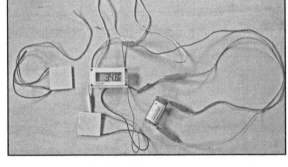

Peltier modules connected to supercapacitor to produce heat and cold.

Motors

Solar powered supercapacitors can provide energy directly to motors for a variety of purposes. One such use is a sun tracking system for a solar panel array. Another use is for well and irrigation water pumps. Fans are also good use of a solar supercapacitor system on hot sunny days.

I use a lot of small hobby motors for different projects and I thought it would be interesting to make a solar supercapacitor powered model plane. Upon

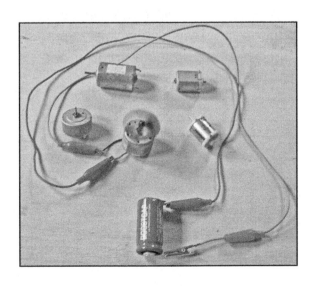

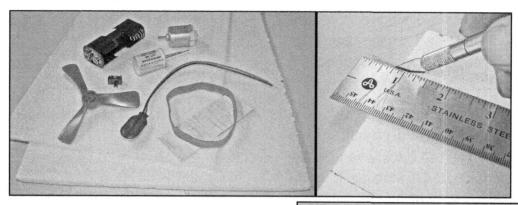

searching the net I found a supercapacitor powered foam plane offered by *Kelvin*. It was fun putting it together. The plane kit comes with flat foam sheets which you have to measure and cut according to the included template. The kit also includes a hobby motor, propeller, supercapacitor, wiring, switch, battery holder and battery clip.

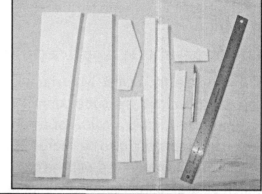

The kit is designed to charge the supercapacitor with a battery but I discarded the battery holder and battery clip and used the 2.5 volt system panel for quick field charging instead. It is a simple affair and held together with glue, tape and rubber bands, but it worked well after a little tweaking, and is fun to fly. They also have a supercapacitor powered foam dragster and hovercraft kit.

A good flying project could be based on *The Foam And Tape Cub Book*. This book shows how to build a Cub Styrofoam plane, which is radio controlled. It would be a challenge to see if you can get it up in the air with a supercapacitor powered motor and control unit.

Some model rocket kits from Estes now have video cameras, which would be an interesting solar supercapacitor project. Basically all you would have to do is modify the rocket to accept the supercapacitor circuit instead of the battery. A supercapacitor would work well for this application because the flight duration is short, and you should be able to get the same run time with a supercapacitor as with a battery. A balloon rocket launching project wherein the balloon lifts the rocket to around 100,000 feet and then launches the rocket would be another good solar supercapacitor candidate to power a number of the circuits. Solar charged supercapacitors can power everything from receivers, transmitters, data sensors, firing mechanisms, video cameras and so on for this sort of project.

Shape memory alloy

Shape memory alloy (also known by other names such as Flexinol, Nitinol, bio-metal, and muscle wire)

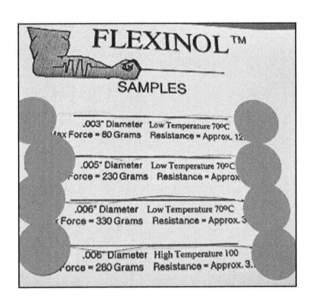

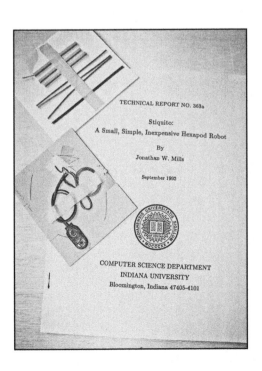

has interesting characteristics wherein it can change its length with the application of an electrical current. This is useful for robots, valves, levers and a myriad of other electro-mechanical devices.

Nitinol or its variants can be found at *McMaster Carr, Dynalloy, Images Scientific Instruments, Mondotronics,* and other suppliers. Kits such as the hexapod robot and other devices using shape memory alloy can easily be powered without batteries by substituting solar supercapacitor power supplies.

Nickel chromium wire

Another material that is interesting in solar supercapacitor powered experiments is nickel chromium wire (Nichrome). This is the same wire used in the rocket ignition system, p.106 in this book. Depending on the length and diameter of the wire, different temperatures can be obtained for a variety of purposes. One use that comes to mind is a foam cutter and shaper, and if you really feel ambitious, you could invent the first solar supercapacitor toaster.

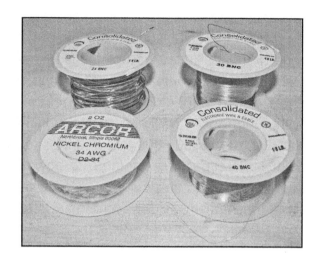

Build Your Own Solar Panel

by Phillip Hurley

Whether you're trying to get off the grid, or you just like to experiment, Build Your Own Solar Panel has all the information you need to build your own photovoltaic panel to generate electricity from the sun. The new revised and expanded edition has easy-to-follow directions, and over 150 detailed photos and illustrations. Materials and tools lists, and links to suppliers of PV cells are included. Every-day tools are all that you need to complete these projects.

Build Your Own Solar Panel will show you how to:

- Design and build PV panels
- Customize panel output
- Make tab and bus ribbon
- Solder cell connections
- Wire a photovoltaic panel

- Purchase solar cells
- Test and rate PV cells
- Repair damaged solar cells
- Work with broken cells
- Encapsulate solar cells

www.buildasolarpanel.com

Solar II

by Phillip Hurley

Now that you've built your solar panels, how do you set up a PV system and plug in? In *Solar II*, Phillip Hurley, author of *Build Your Own Solar Panel* will show you how to:

- Plan and size your solar electric system

- Build racks and charge controllers

- Mount and orient PV panels

- Wire solar panel arrays

- Make a ventilated battery box

- Wire battery arrays for solar panels

- Install an inverter

- Maintain solar batteries for optimum life and performance

- Make your own combiner box, bus bars and DC service box

www.buildasolarpanel.com

The Battery Builder's Guide
by Phillip Hurley

The Battery Builder's Guide is a practical hands-on text that will show you how to make your own rechargeable flooded lead acid batteries. Learn how to recycle parts and materials, how to fabricate battery components and where to purchase the parts, materials and tools you need to build or rebuild batteries. The text covers construction of batteries with Plante (pure lead) and Faure (pasted lead) plates.

Topics include:

- Recycling old lead acid batteries
- Molding battery parts
- Design formulas and tables
- Lead burning
- Techniques and tools for battery building
- Building plate burning racks
- Pasting and forming plates
- Types of batteries such as SLA and deep cycle, and their characteristics and uses
- And more... all illustrated with extensive step-by-step photos

Flooded lead acid batteries are used for stationary applications such as solar and wind powered electrical systems, and for mobile applications. If you need custom batteries of a specific size or output, wish to experiment with building batteries, or want to lower your costs by using recycled components and materials, *The Battery Builder's Guide* has the information you need.

www.batterybuildersguide.com

Build Your Own Fuel Cells

by Phillip Hurley

The technology of the future is here today - and now available to the non-engineer! *Build Your Own Fuel Cells* contains complete, easy to understand illustrated instructions for building several types of proton exchange membrane (PEM) fuel cells - and, templates for 6 PEM fuel cell types, including convection fuel cells and oxygen-hydrogen fuel cells, in both single slice and stacks.

Low tech/high quality

Two different low-tech fuel cell construction methods are covered: one requires a bandsaw and drill press, and the other only a few hand tools. Anyone with minimum skills and tools will be able to produce high quality fuel cells from readily obtainable materials - contact info for materials suppliers is included.

Electrolyzers and MEAs

Build Your Own Fuel Cells includes a detailed discussion of building a lab electrolyzer to generate hydrogen to run fuel cells - and templates for the electrolyzer. Also covered is setting up a PV solar panel to power the electrolyzer, and experimental low-tech methods for producing membrane electrode assemblies (MEAs - the heart of the fuel cell).

Build Your Own Fuel Cells, 221 pages, over 140 B&W photos and illustrations, including 39 templates.

www.buildafuelcell.com

Build A Solar Hydrogen Fuel Cell System

by Phillip Hurley

Learn how to construct and operate the components of a solar hydrogen fuel cell system: the fuel cell stack, the electrolyzer to generate hydrogen fuel, simple hydrogen storage, and solar panels designed specifically to run electrolyzers for hydrogen production. Complete, clear, illustrated instructions to build all the major components make it easy for the non-engineer to understand and work with this important new technology.

Featured are the author's innovative and practical designs for efficient solar powered hydrogen production including:

◆ ESPMs (Electrolyzer Specific Photovoltaic Modules) – 40 watt solar panels designed specifically to run electrolyzers efficiently;

◆ a 40-80 watt electrolyzer for intermittant power from renewable energy sources such as solar and wind;

◆ and, a 6-12 watt planar hydrogen fuel cell stack to generate electricity.

Any of these components can be ganged or racked, or scaled up in size for higher output. You'll also learn how to set up an entire gas processing system, and where to find parts and materials – everything you need for an experimental stationary unit that will give you a solid base for building and operating systems for larger power needs. There are even schematics for adapting conventional solar panels (BSPMs – Battery Specific Photovoltaic Modules) for efficient hydrogen production, and setting up hybrid (battery and fuel cell) PV systems.

www.solarh.com

Practical Hydrogen Systems:
an Experimenter's Guide
by Phillip Hurley

The author of *Build Your Own Fuel Cells, Build Your Own Solar Panel* and *Build a Solar Hydrogen Fuel Cell System* shares the details of his high quality, self-pressurized stainless steel hydrogen generator, and offers valuable insight into components, subsystems and processes involved in the production of hydrogen.

Hurley's innovative small scale system is designed for ease of fabrication and a minimum of components - in versatile modular configurations that make experimenting and customizing easy. This hydrogen-generating system features automatic pressure control, and the book includes complete schematics for the electronic/electromechanical process controller, as well as over 230 other illustrations.

Topics covered include:

- Electrolyzer construction and design
- Hydride storage
- Cylinder hydrogen
- Hydrogen leak detection and detectors
- Catalytic recombiners
- Hydrogen processing systems
- Filtration
- Parts suppliers and resources for hydrogen system components
- Fuel cell units
- Hydrogen storage
- Bubblers and scrubbing
- Pressure vessels for hydrogen systems
- Tubing and fittings for hydrogen systems

- Electrolytes
- Nitrogen purging for hydrogen systems
- Hydrogen regulators
- Material compatability for hydrogen system components
- Electronic and electromechanical process controllers for hydrogen systems
- Meters and pressure gauges for hydrogen process measurement
- Pressure switches and level switches for electrolyzers and hydrogen storage
- Intrinsically safe systems for hydrogen production
- Pressurized safety systems for hydrogen process control equipment

www.solarh.com

Made in the USA
Las Vegas, NV
22 July 2021